Safari® WebKit Development for iPhone® OS 3.0

Safari® WebKit Development for iPhone® OS 3.0

Safari® WebKit Development for iPhone® OS 3.0

Richard Wagner

Wiley Publishing, Inc.

Safari® WebKit Development for iPhone® OS 3.0

Published by
Wiley Publishing, Inc.
10475 Crosspoint Boulevard
Indianapolis, IN 46256
www.wiley.com

To KimmyWags and the J-Team

About the Author

Richard Wagner is Senior Developer at Maark, LLC, and author of *Professional iPhone and iPod Touch Programming: Building Applications for Mobile Safari* and several Web-related books on the underlying technologies of the iPhone application platform. These books include *Creating Web Pages All-In-One Desk Reference For Dummies, XML All-In-One Desk Reference For Dummies, XSLT For Dummies, Web Design Before & After Makeovers,* and *JavaScript Unleashed* (1st, 2nd ed.). Previously, Richard was Vice President of Product Development at NetObjects and chief architect of a CNET award-winning JavaScript development tool named NetObjects ScriptBuilder.

Credits

Executive Editor
Carol Long

Project Editor
Kelly Talbot

Technical Editor
Michael Morrison

Production Editor
Rebecca Anderson

Copy Editor
Karen Gill

Editorial Director
Robyn B. Siesky

Editorial Manager
Mary Beth Wakefield

Marketing Manager
David Mayhew

Production Manager
Tim Tate

Vice President and Executive Group Publisher
Richard Swadley

Vice President and Executive Publisher
Barry Pruett

Associate Publisher
Jim Minatel

Project Coordinator, Cover
Lynsey Stanford

Compositor
Jeff Lytle, Happenstance Type-O-Rama

Proofreader
Josh Chase, Word One

Indexer
Robert Swanson

Cover Image
© istockphoto/jabiru

Cover Designer
Michael E. Trent

Acknowledgments

In the three years since its release, the iPhone has grown to be my favorite piece of technology I have ever owned. As such, the topic of iPhone Web application development has been a joy to write about. However, the book was also a joy because of the stellar team I had working with me on this book. First and foremost, I'd like to thank Kelly Talbot for his masterful role as project editor. He kept the project on track and running smoothly from start to finish. I'd also like to thank Michael Morrison for his insights and ever-watchful eye that ensured technical accuracy in this book. Further, thanks also to Karen Gill for her editing prowess.

Finally, I'd be remiss not to offer a deep "thank you" to Carol Long and Matt Wagner for their roles in getting this book off the ground.

Contents

Contents

Chapter 5: Styling with CSS — 119

Chapter 6: Programming the Interface — 135

Chapter 7: Handling Touch Interactions and Events — 185

Contents

Contents

Introduction

The amazing success of the iPhone over the past two years clearly indicates that application developers are entering a brave new world of sophisticated, multifunctional mobile applications. No longer do applications and various media need to live in separate silos. Instead, mobile Web-based applications can bring together elements of Web 2.0 apps, traditional desktop apps, multimedia video and audio, and cell phones.

This book covers the various aspects of developing Web-based applications for the iPhone and iPod touch. Specifically, you will discover how to create a mobile application from the ground up, utilize existing open source frameworks to speed up your development times, emulate the look and feel of built-in Apple applications, capture finger touch interactions, use Ajax to load external pages, and optimize applications for WiFi and 3G networks.

Who This Book Is For

This book is aimed primarily for Web developers already experienced in Web technologies who want to build new applications for iPhone or migrate existing Web apps to this platform. In general, readers should have a working knowledge of the following technologies:

- HTML/XHTML
- CSS
- JavaScript
- Ajax

However, if you are less experienced working with these technologies, be sure to take advantage of the primer at the start of this book.

What This Book Covers

This book introduces readers to the Web application platform for iPhone OS 3.0. It guides readers through the process of building new applications from scratch and migrating existing Web applications to this new mobile platform. As it does so, it helps readers design a user interface that is optimized for the iPhone touch-screen display and integrate their applications with iPhone services, including Phone, Mail, Google Maps, and GPS.

Introduction

The chapter-level breakdown is as follows:

1. **Introducing Safari/WebKit Development for iPhone 3.0:** Explores the Safari development platform and walks you through the ways you can develop for iPhone.

2. **Working with Core Technologies:** Provides an overview of some of the key technologies you'll be working with as you develop iPhone Web apps. Pays particular attention to scripting and the Document Object Model.

3. **Building with Web App Frameworks:** Highlights the major open source iPhone Web app frameworks and shows you how to be productive quickly with each of them.

4. **Designing a Usable and Navigable UI:** Overviews the key design concepts and principles you need to use when developing a highly usable interface for Safari on iPhone.

5. **Styling with CSS:** Discusses WebKit-specific styles that are useful for developing Web apps for iPhone and iPod touch.

6. **Programming the Interface:** Provides a code-level look at developing an iPhone Web application interface.

7. **Handling Touch Interactions and Events:** The heart of an iPhone is its touch screen interface. This chapter explores how to handle touch interactions and capture JavaScript events.

8. **Programming the Canvas:** Safari on iPhone browser provides full support for canvas drawing and painting, opening opportunities for developers.

9. **Special Effects and Advanced Graphics:** The Safari canvas provides an ideal environment for advanced graphics techniques, including gradients and masks.

10. **Integrating with iPhone Services:** Discusses how a Web application can integrate with core iPhone services, including Phone, Mail, Google Maps, and GPS.

11. **Offline Applications:** Covers how you can use HTML 5 offline cache to create local Web apps that don't need a live server connection.

12. **Enabling and Optimizing Web Sites for the iPhone and iPod Touch:** Covers how to make an existing Web site compatible with Safari and then how to optimize the site for use as a full-fledged Web application.

13. **Bandwidth and Performance Optimizations:** Deals with the all-important issue of performance of Web-based applications and what techniques developers can use to minimize constraints and maximize bandwidth and app execution performance.

14. **Packaging Apps as Bookmarks: Bookmarklets and Data URLs:** This chapter explains how you can employ two little-used Web technologies to support limited offline support.

15. **Debug and Deploy:** Discusses various methods of debugging Safari Web applications.

16. **iPhone SDK: From Web App to Native App:** How do you know when you need to move your Web app to a native iPhone? This chapter explores migration strategies and introduces you to the iPhone SDK.

What You Need to Use This Book

To work with the examples of this book, you need the following:

- iPhone
- Safari for Mac or Windows

The complete source code for the examples is available for download from our Web site at `www.wrox.com`.

Conventions

I used several conventions throughout this book to help you get the most from the text.

- New terms are *italicized* when I introduce them.
- URLs and code within the text are in a monospaced font, such as `<div class="panel">`.
- Within blocks of source code, I occasionally want to highlight a specific section of the code. To do so, I use a gray background. For example:

```
addEventListener("load", function(event) {
    convertSrcToImage(0);
    photoEnabled = true;
    showPhoto(1);
    }, false);
```

Source Code

As you work through the examples in this book, you can type in all the code manually or download the source code files from the Wrox Web site (`www.wrox.com`). At the site, locate the book's detail page using Search or by browsing through the title listings. On the page, click the Download Code link, and you are ready to go.

You may find it easiest to search by ISBN number. This book's ISBN is 978-0-470-54966-7.

Errata

The editors and I worked hard to ensure that the contents of this book are accurate and there no errors either in the text or in the code examples. However, in case future iPhone OS releases impact what's been said here, I recommend visiting `www.wrox.com` and checking out the Book Errata link. You will be taken to a page that lists all errata that has been submitted for the book and posted by Wrox editors.

If you discover an issue that is not found on the Errata page, I would be grateful if you let us know about it. To do so, go to `www.wrox.com/contact/techsupport.shtml` and provide this information in the online form. The Wrox team will double-check your information and, as appropriate, post it on the Errata page. In addition, Wrox will correct the problem in future versions of the book.

1

Introducing Safari/WebKit Development for iPhone 3.0

The introduction of the iPhone and the subsequent unveiling of the iPod touch have revolutionized the way people interact with handheld devices. No longer do users have to use a keypad for screen navigation or browse the Web through "dumbed down" pages. These mobile devices have brought touch screen input, a revolutionary interface design, and a fully functional Web browser right into the palms of people's hands.

Seeing the platform's potential, all the segments of the developer community jumped on board. Although native applications may receive most of the attention, you can still create apps for iPhone without writing a single line of Objective-C, the programming language used to develop native iPhone apps. In fact, iPhone's WebKit-based browser provides a compelling application development platform for Web developers who want to create custom apps for iPhone using familiar Web technologies.

Each subsequent release of the iPhone OS and Safari on iPhone has put increased power into the hands of Web developers, and as I'll discuss shortly, the iPhone OS 3.0 release is no exception.

Discovering the Safari/WebKit Platform

An iPhone Web application runs inside of the built-in Safari browser that is based on Web standards, including these:

- ❑ HTML/XHTML (HTML 4.01 and XHTML 1.9, XHTML mobile profile document types)
- ❑ CSS (CSS 2.1 and partial CSS3)

- ❑ JavaScript (ECMAScript 3, JavaScript 1.4)
- ❑ AJAX (for example, XMLHTTPRequest)
- ❑ SVG (Scalable Vector Graphics) 1.1
- ❑ HTML 5 media tags
- ❑ Ancillary technologies (video and audio media, PDF, and so on)

Safari on iPhone and iPod touch (which I refer to throughout the book as simply *Safari*) becomes the platform upon which you develop applications and becomes the shell in which your apps must operate (see Figure 1-1).

Figure 1-1: Safari user interface

Safari is built with the same open source WebKit browser engine as Safari for OS X and Safari for Windows. However, although the Safari family of browsers is built on a common framework, you'll find it helpful to think of Safari on iPhone as a close sibling to its Mac and Windows counterparts, not an identical twin to either of them. Safari on iPhone, for example, does not provide the full extent of CSS or JavaScript functionality that its desktop counterpart does.

In addition, Safari on iPhone provides only a limited number of settings that users can configure. As Figure 1-2 shows, users can turn off and on support for JavaScript, plug-ins, and a pop-up blocker. Users can also choose whether they want to always accept cookies, accept cookies only from sites they visit, or never accept cookies. A user can also manually clear the history, cookies, and cache from this screen.

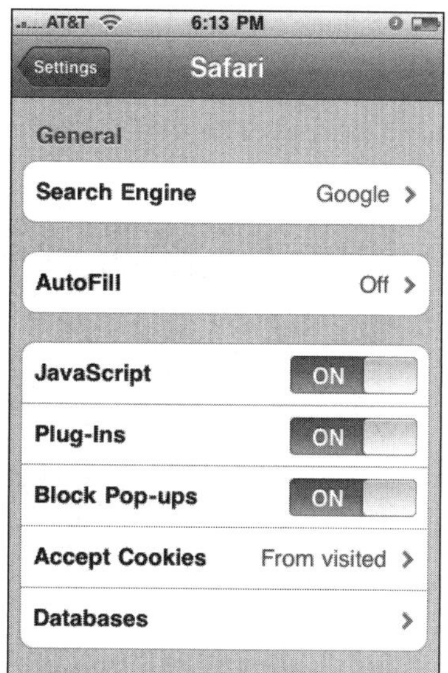

Figure 1-2: Safari on iPhone preferences

Quite obviously, native apps and Web apps are not identical — both from developer and end-user standpoints. From a developer standpoint, the major difference is the programming language — utilizing Web technologies rather than Objective-C. However, there are also key end-user implications, including these:

❑ **Performance:** The performance of a Safari-based Web application is not going to be as responsive as a native compiled application, both because of the interpretive nature of web scripting as well as the fact that the application operates over Wi-Fi and 3G networks. (Remember, iPod touch supports Wi-Fi access only.) However, in spite of the technological constraints, you can perform many optimizations to achieve acceptable performance. (Several of these techniques are covered in Chapter 13, "Bandwidth and Performance Optimizations.")

Table 1-1 shows the bandwidth performance of Wi-Fi, 3G, and the older EDGE networks.

Table 1-1: Network Performance

Network	Bandwidth
Wi-Fi	54 Mbps
3G	Up to 7.2 Mbps
EDGE	70–135 Kbps, 200 Kbps burst

❏ **Launching:** All native applications are launched from the main Home screen of the iPhone and iPod touch (see Figure 1-3). In the original iPhone OS release, Apple provided no way for Web apps to be launched from here, requiring Web apps to be accessed from the Safari Bookmarks list. Fortunately, subsequent releases of the iPhone OS have given users the ability to add "Web Clip" icons for their Web apps (such as the Cup-O-Joe Web app in Figure 1-4).

Figure 1-3: Built-in applications launch
from the main Home screen

Figure 1-4: Web applications can also
be included on the Home screen

❑ **User interface (UI):** Native iPhone applications often adhere to Apple UI design guidelines. Fortunately, using open source frameworks and standard Web technologies, you can closely emulate native application design using a combination of HTML, CSS, and JavaScript. Figures 1-5 and 1-6 compare the UI design of a native application and a Safari-based Web application.

What's more, iPhone OS enables you to hide all Safari browser UI elements through meta tags, enabling you to essentially emulate the look and feel of a native app. (See Figure 1-7.)

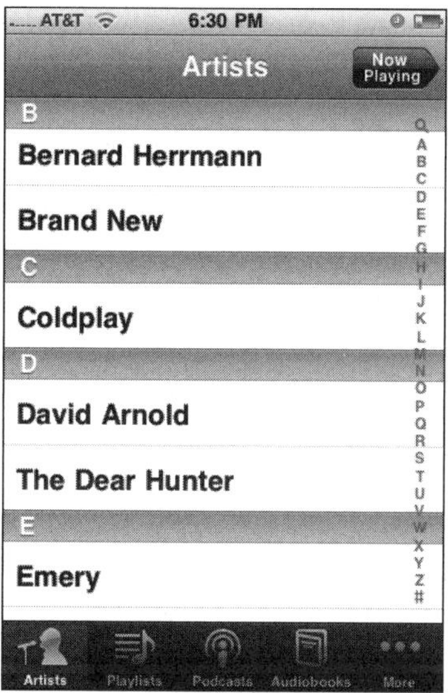

Figure 1-5: Edge-to-edge navigation pane in the native app

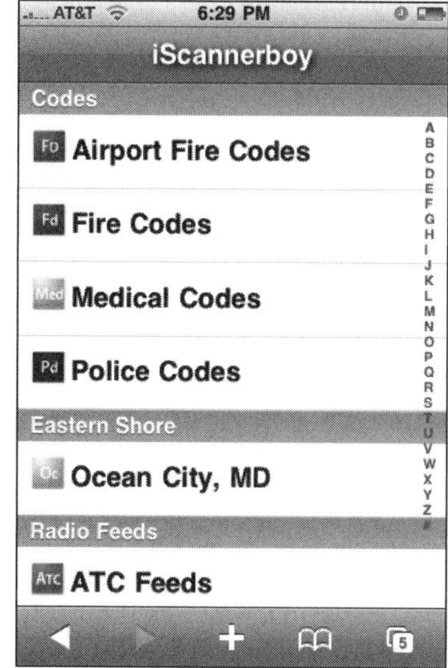

Figure 1-6: Edge-to-edge navigation pane in a custom Web application

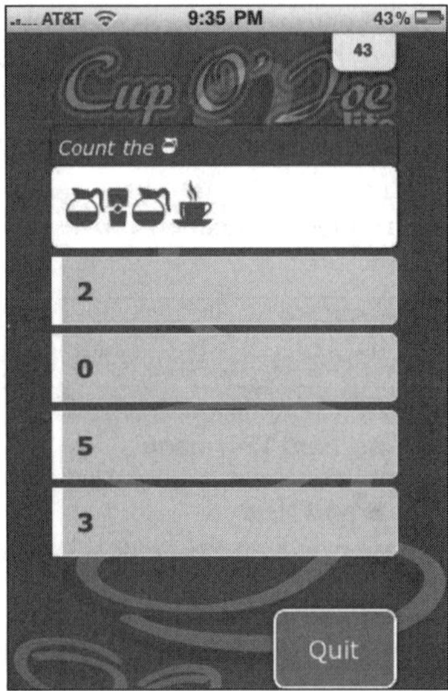

Figure 1-7: Web app or native app?
It's hard to tell.

What's New in iPhone OS 3.0 for Web App Developers

There are several new capabilities available to Web app developers with the release of iPhone OS 3.0 and Safari 4.0 iPhone. These are highlighted here:

❑ **Geolocation:** Safari on iPhone now supports HTML 5 geolocation capabilities, which enable JavaScript to interact with iPhone's GPS service to retrieve the current location of the iPhone (see Figures 1-8 and 1-9). As a result, you can create apps that can broadcast the location of a GPS-enabled iPhone.

Google is using this capability with its Latitude service for sharing your location with your friends.

❑ **HTML 5 Media Tags:** The newest release of Safari on iPhone supports HTML 5 `video` and `audio` tags for embedding video and audio content in Web pages. These new elements eliminate the need for complicated `embed` and `object` tags for embedding multimedia elements and allow you to utilize a powerful JavaScript API. What's more, because iPhone doesn't support Flash, you can use the `video` tag to embed QuickTime MOV files.

Safari is the first major browser to provide full support for HTML 5 media tags; therefore, you have to be careful in their usage on standard Web sites because other browsers may not support it yet. However, because you are creating an app specifically for the iPhone, you can make full use of these tags.

❑ **CSS animation and effects:** The new release of Safari supports *CSS animation*, which enables you to manipulate elements in various ways, such as scaling, rotating, fading, and skewing. Safari on iPhone also supports *CSS effects*, which enable you to create gradients, masks, and reflections entirely through CSS.

❑   **SVG:** SVG (or Scalable Vector Graphics) is a new XML-based format for creating static and animated vector graphics. With SVG support, Safari on iPhone not only provides a way to work with scalable graphics, but actually provides a technology that could replace the need for Flash to create animated media.

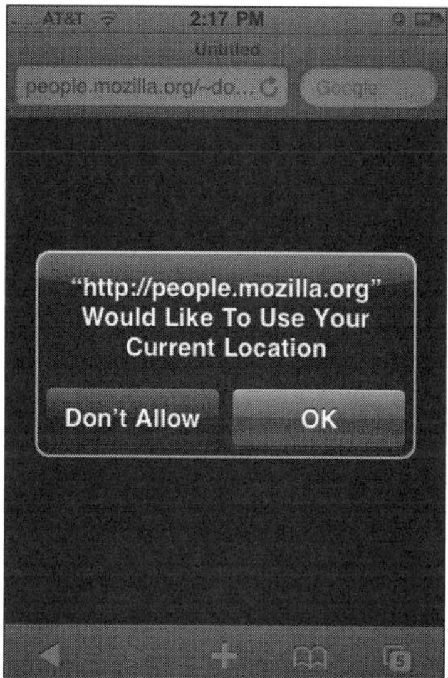

Figure 1-8: Users are asked to confirm GPS location services support

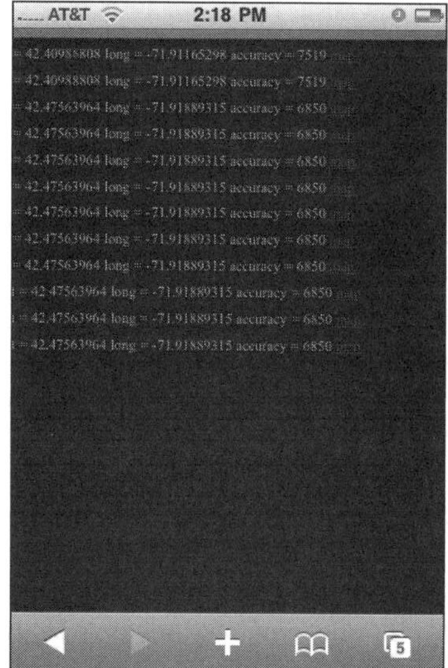

Figure 1-9: Test Web app that displays geolocation info in real time

Four Ways to Develop Web Apps for iPhone

A Web application that you can run in any browser and an iPhone Web application are certainly made using the same common ingredients — HTML, CSS, JavaScript, and AJAX — but they are not identical. In fact, there are four approaches to consider when developing for iPhone:

❑ **Level 1 — Fully compatible Web site/application:** The ground-level approach is to develop a Web site/app that is "iPhone/iPod touch-friendly" and is fully compatible with the Apple mobile devices (see Figure 1-10). These sites avoid using technologies that the Apple mobile devices do not support, including Flash, Java, and other plug-ins. The basic structure of the presentation layer also maximizes the use of blocks and columns to make it easy for users to navigate and zoom within the site. This basic approach does not do anything specific for iPhone users but makes sure that there are no barriers to a satisfactory browsing experience. (See Chapter 12, "Enabling and Optimizing Web Sites for the iPhone and iPod Touch," for converting a Web site to be friendly for iPhone users.)

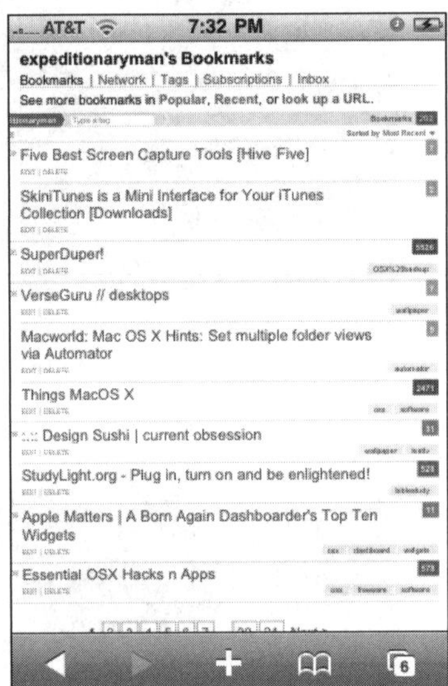

Figure 1-10: The site is easy to navigate

❑ **Level 2 — Web site/application optimized for Safari:** The second level of support for iPhone is to not only provide a basic level of experience for the Safari on iPhone user, but to provide an optimized experience for those who use Safari browsers, such as utilizing some of the enhanced WebKit CSS properties supported by Safari.

❑ **Level 3 — Dedicated iPhone/iPod touch Web site/application:** A third level of support is to provide a Web site tailored to the viewport dimensions of the iPhone and provide a strong Web browsing experience for Apple device users (see Figures 1-11 and 1-12). However, although these sites are tailored for iPhone viewing, they do not always seek to emulate Apple UI design. And, in many cases, these are often stripped-down versions of a fuller Web site or application.

Figure 1-11: Amazon's mobile site

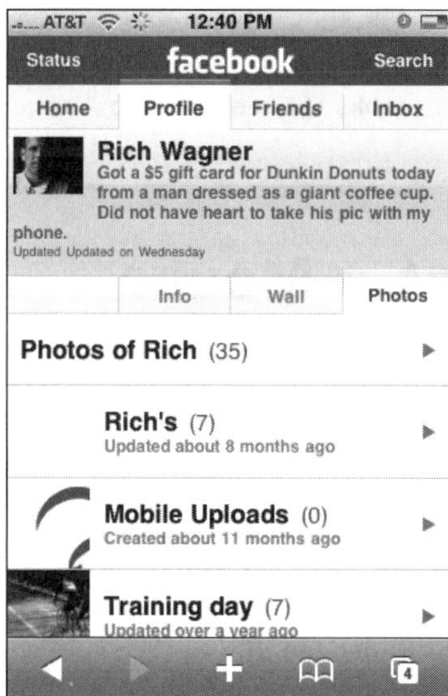

Figure 1-12: Facebook's dedicated
site for iPhone

❏ **Level 4 — Native-looking iPhone Web application:** The final approach is to provide a Web application that is designed exclusively for iPhone and closely emulates the UI design of native applications (see Figure 1-13). One of the design goals is to minimize users' awareness that they are even inside of a browser environment. Moreover, a full-fledged iPhone application will, as is relevant, integrate with iPhone-specific services, including Phone, Mail, and Google Maps.

Therefore, as you consider your application specifications, be sure to identify which level of user experience you want to provide iPhone users, and design your application accordingly. This book focuses primarily on developing native-looking Web applications.

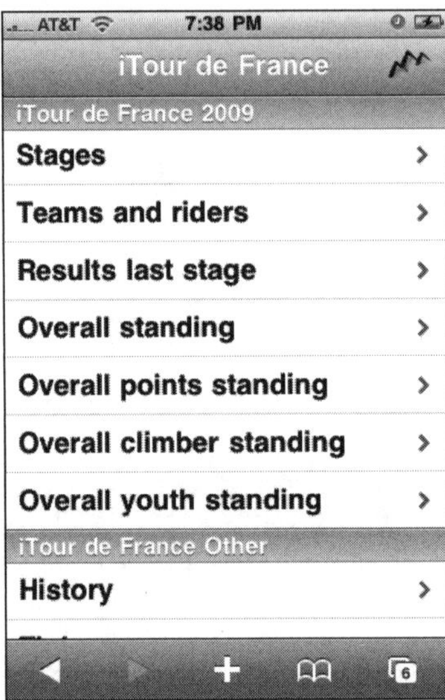

Figure 1-13: iPhone Web app that looks like a native app

The Finger Is Not a Mouse

As you develop applications for iPhone, one key design consideration that you need to drill into your consciousness is that *the finger is not a mouse*. On the desktop, a user can use a variety of input devices — such as an Apple Mighty Mouse, a Logitech trackball, or a laptop touchpad. But, on-screen, the mouse pointer for each of these pieces of hardware is always identical in shape, size, and behavior. However, on iPhone, the pointing device is always going to be unique. Ballerinas, for example, will probably input with tiny, thin fingers, whereas NFL players will use big, fat input devices. Most of the rest of us will fall somewhere in between. Additionally, fingers are not nearly as precise as mouse pointers are, making interface sizing and positioning issues very important, whether you are creating an iPhone-friendly Web site or a full-fledged iPhone Web application.

Also, finger input does not always correspond to mouse input. A mouse has a left-click, right-click, scroll, and mouse move. In contrast, a finger has a tap, flick, drag, and pinch. However, as an application developer, you will want to manage what types of gestures your application supports. Some of the gestures that are used for browsing Web sites (such as the double-tap zoom) are actually not something you want to support inside of an iPhone Web app. Table 1-2 displays the gestures that are supported on iPhone as well as whether this type of gesture is typically supported on a Web site or a full Web application.

Table 1-2: Finger Gestures

Gesture	Result	Web site	App
Tap	Equivalent to a mouse click	Yes	Yes
Drag	Moves around the viewport	Yes	Yes
Flick	Scrolls up and down a page or list	Yes	Yes
Double-tap	Zooms in and centers a block of content	Yes	No
Pinch open	Zooms in on content	Yes	No
Pinch close	Zooms out to display more of a page	Yes	No
Touch and hold	Displays an info bubble	Yes	No
Two-finger scroll	Scrolls up and down an `iframe` or element with the CSS `overflow:auto` property	Yes	Yes

Limitations and Constraints

Because iPhone is a mobile device, it is obviously going to have resource constraints that you need to be fully aware of as you develop applications. Table 1-3 lists the resource limitations and technical constraints. What's more, certain technologies (listed in Table 1-4) are unsupported, and you need to steer away from them when you develop for iPhone and iPod touch.

Table 1-3: Resource Constraints

Resource	Limitation
Downloaded text resource (HTML, CSS, JavaScript files)	10MB
JPEG images	128MB (all JPEG images over 2MB are subsampled — decoding the image to 16x fewer pixels)
PNG, GIF, and TIFF images	8MB (in other words, width*height*4<8MB)
Animated GIFs	Less than 2MB ensures that frame rate is maintained (over 2MB, only first frame is displayed)
Nonstreamed media files	10MB
PDF, Word, Excel documents	30MB and up (very slow)
JavaScript stack and object allocation	10MB
JavaScript execution limit	5 seconds for each top-level entry point (`catch` is called after 5 seconds in a `try/catch` block)
Open pages in Mobile Safari	8 pages

Table 1-4: Technologies Not Supported by iPhone and iPod touch

Area	Technologies not supported
Web technologies	Flash media, Java applets, SOAP, XSLT, and plug-in installation
Mobile technologies	WML
File access	Local file system access
Security	Diffie-Hellman protocol, DSA keys, self-signed certificates, and custom x.509 certificates
JavaScript events	Several mouse-related events (see Chapter 7, "Handling Touch Interactions and Events")
JavaScript commands	`showModalDialog()`, `print()`
Bookmark icons	`ICO` files
HTML	`Input type="file"`, tool tips
CSS	Hover styles, `position:fixed`

Setting Up Your Development Environment on a Local Network

Because iPhone does not allow you to access the local file system, you cannot place your application directly onto the device. As a result, you need to access your Web application through another computer. On a live application, you will obviously want to place your application on a publicly accessible Web server. However, testing is another matter. If you have a Wi-Fi network at your office or home, I recommend running a Web server on your main desktop computer to use as your test server during deployment.

If you are running Mac OS X, you already have Apache Web server installed on your system. To enable iPhone access, go to System Preferences Sharing Services and turn the Personal Web Sharing option on (see Figure 1-14). When this feature is enabled, the URL for the Web site is shown at the bottom of the window. You'll use this base URL to access your Web files from your iPhone or iPod touch.

You can add files either in the computer's Web site directory (`/Library/WebServer/Documents`) or your personal Web site directory (`/Users/YourName/Sites`) and then access them from the URL bar on your iPhone (see Figure 1-15).

If your users experience crashing or instability inside Safari, direct them to clear the cache by clicking the Clear Cache button in the Safari Settings pane.

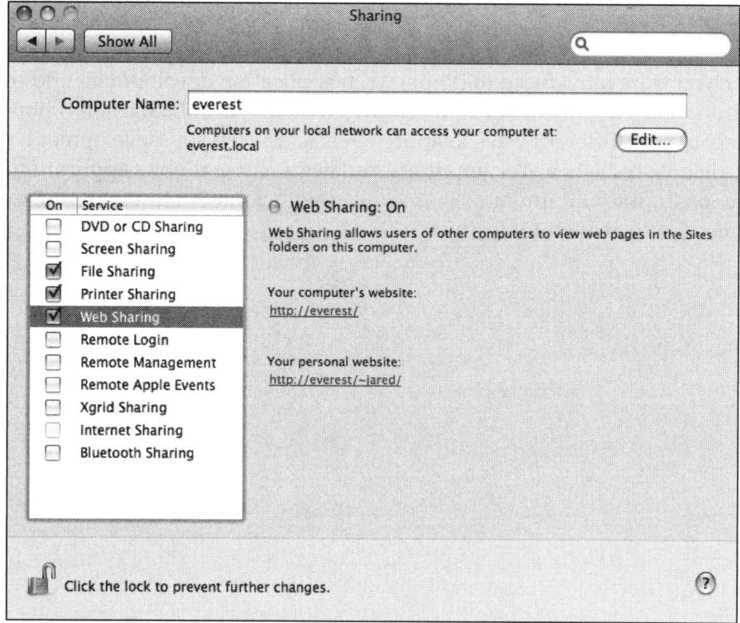

Figure 1-14: Turn on Personal Web Sharing

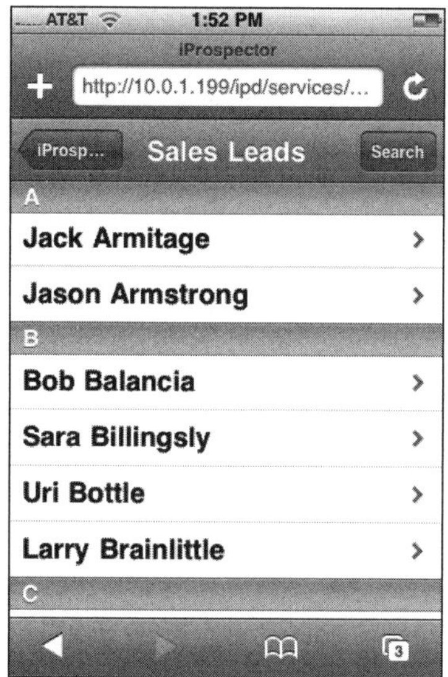

Figure 1-15: Accessing desktop files
from an iPhone

Summary

In this chapter, you were introduced to iPhone Web application development and the Safari/WebKit browser. It began with a survey of key new features that are part of Safari on iPhone for iPhone OS 3.0. I then talked about four different ways to approach iPhone Web app development, ranging from an iPhone-compatible Web site to a Web app that emulates a native iPhone application. Then, after a discussion on the constraints and limitations associated with Safari, I closed by showing you how to set up your development environment for testing.

2

Working with Core Technologies

Whereas a native iPhone app is built entirely in Objective-C, an iPhone Web app is composed of a variety of core technologies that serve as interlocking building blocks. HTML provides structure for the user interface and application data, and CSS is used for styling and presentation. JavaScript and Ajax provide the programming logic used to power the app. And, depending on the app, it may have a backend application server, such as Java or PHP.

Books on programming native iPhone apps often have a primer chapter on the basics of Objective-C to make sure everyone is speaking the same language, so to speak. And, although an entire book could easily be filled writing a primer on the core Web technologies you will work with to develop an iPhone Web app, I do want to explore some of the key technologies you'll need to be sure you know about to be successful.

So as not to start from ground zero, I will assume that you at least know the basics of HTML and have a working knowledge of CSS. That's why I will spend most of my time talking about both the scripting logic layer and the Document Object Model (DOM). However, before I do so, I want to highlight the new HTML 5 tags that Safari on iPhone supports for embedding media into your Web app.

Exploring HTML 5 Media Elements

Before iPhone OS 3.0, working with video inside of an iPhone Web app usually consisted of a simple link to a YouTube clip, which then launched the YouTube app. Because Safari on iPhone didn't (and still doesn't) support Flash video (FLV), there were few alternatives to an ordinary link when working with video. However, one of the key HTML 5 technologies that Safari on iPhone now supports is the `video` element.

The `video` element defines a video clip or stream, much like an `img` tag defines an image on your page. The promise of the `video` tag is that it eliminates the complicated hoops that developers have to go through now with embedded media content. Instead of mixing complicated `object` definitions and script, you can embed media with a simple tag definition.

Unfortunately for normal Web sites, the `video` element remains something of a tease, because most desktop browsers don't yet support HTML 5. As a result, developers have to either add code for unsupported browsers or avoid its use altogether.

However, if you are creating an iPhone Web app, you don't have this same dilemma. Safari on iPhone, starting with the OS 3.0 release, provides full support. Therefore, if you need to utilize video in your app, be sure to take advantage of the `video` tag.

Note that the video does not play inside the Web page as an embedded video, but instead launches the built-in iPhone media player, which occupies the full screen of the iPhone. The user then clicks the Done button to return to your app.

The basic syntax for the element is shown here:

```
<video src="../video/trailer.mov" controls="true" poster="picture.jpg" width="300"
height="200"/>
```

Table 2-1: Attributes for video Element

Attribute	Description
autoplay	When set to `true`, the video plays as soon as it is ready to play.
controls	If `true`, the user is shown playback controls.
end	Specifies the endpoint to stop playing the video. If it's not defined, the video plays to the end.
height	Defines the height of the video player.
loopend	Defines the ending point of a loop.
loopstart	Defines the starting point of a loop.
playcount	Specifies the number of times a video clip is played. Defaults to 1.
poster	Specifies the URL of a "poster image" to show before the video begins playing.
src	Defines the URL of the video.
start	Sets the point at which the video begins to play. If `start` is not defined, the video starts playing at the beginning.
width	Defines the width of the video player.

Supported video formats include QuickTime (MOV) and MPEG (MP4). Note that Safari on iPhone does not support Flash media (FLV) or Ogg Theora (OGG).

Adding a video into your app becomes as easy as adding the `video` tag to your page. For example:

```
<!DOCTYPE html PUBLIC "-//W3C//DTD XHTML 1.0 Strict//EN"
          "http://www.w3.org/TR/xhtml1/DTD/xhtml1-strict.dtd">

<html xmlns="http://www.w3.org/1999/xhtml">
<head>
<title>Video</title>
<meta name="viewport" content="width=320; initial-scale=1.0; maximum-scale=1.0;
user-scalable=0;"/>
<style type="text/css">

body
{
  background-color: #080808;
  margin: 10;
}


p
{
  color: #ffffff;
  font-family: Helvetica, sans-serif;
  font-size:10px;
}

</style>

</head>
<body>
  <div style="text-align:center">
  <p>Check out the new trailer for our upcoming video game release.</p>
  <video src="../videos/tlr2_h.640.mov" controls="true" width="300"/>
  </div>
</body>
</html>
```

Figure 2-1 shows how Safari displays the video player. When the user touches the video player, Safari displays the video in the built-in media player, as shown in Figure 2-2.

The `audio` element works in much the same way, although its attributes are a subset of the `video` tag's set. They include `src`, `autobuffer`, `autoplay`, `loop`, and `controls`. Figure 2-3 shows the audio file being played in the media player.

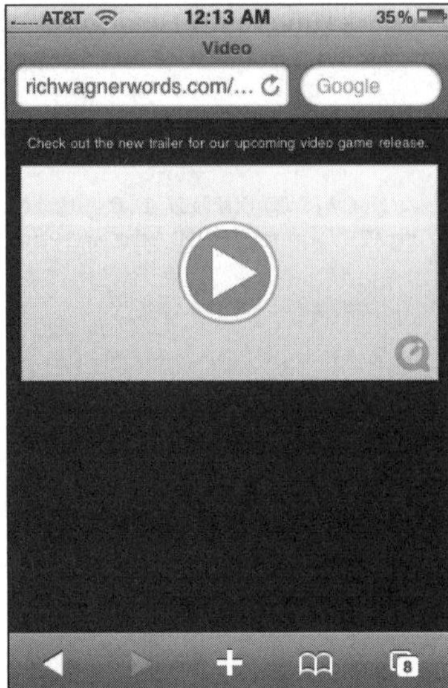

Figure 2-1: Safari displaying the `video` element

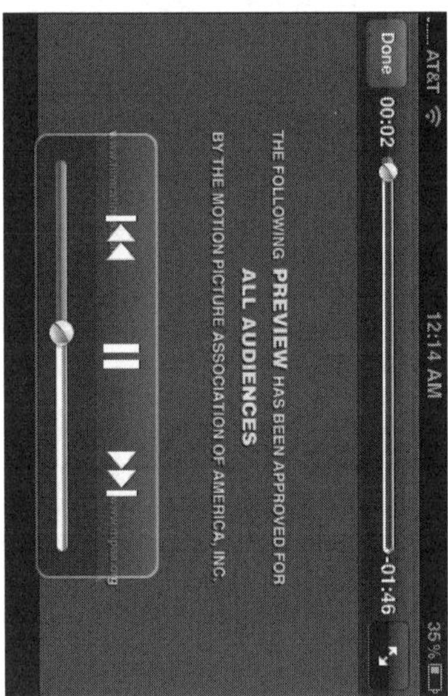

Figure 2-2: Video plays in a separate media player.

Figure 2-3: Audio files are also played
in the media player.

Scripting JavaScript

The scripting layer of your iPhone Web app is often the "brains behind the operation," the place on the
client side in which you add your programming logic. That's why as you begin to develop more and
more sophisticated Web apps for the iPhone platform, you'll find that the technology you'll rely more
and more on is JavaScript. Therefore, before you go any further in this book, you need a solid grasp of
how to script with JavaScript. Use this extended section as a primer for the core essentials of programming with JavaScript.

Syntax and Basic Rules

Like any other programming language, JavaScript has its own syntax rules and conventions. Consider
the following script:

```
<script language="javascript" type="text/javascript">

// Defining a variable
var message = "Hello";

// Creating a function, a module of code
function showGreeting(allow)
{
  // If allow is true, then show the message variable
  if (allow == true)
```

```
  {
    alert(message);
  }
  // Otherwise, show this message
  else
    alert("I withhold all greetings.");
}

// Call function
showGreeting(true)

</script>
```

Notice a few aspects of the JavaScript syntax:

❑   **Case sensitivity:** JavaScript is a case-sensitive language. The core language words (such as `var`, `function`, and `if` in the example) are all lowercase. Built-in JavaScript and Document Object Model objects use lowerCamelCase conventions, which are explained in the later section titled "JavaScript Naming Conventions." Variable names that you define can be in the case you choose, but note that the following terms are not equivalent:

```
var message = "Hello";
var MESSAGE = "Hello";
var Message = "Hello";
```

❑   **Whitespace:** Just like HTML, JavaScript ignores spaces, tabs, and new lines in code statements. Each of the following is equivalent:

```
var message="Hello";
var message = "Hello";
```

These are as well:

```
if (allow==true)
if ( allow == true )
```

❑   **Semicolons:** Each JavaScript statement ends with a semicolon (except for the descriptive comment lines).

❑   **Curley brackets:** Some JavaScript control structures, such as functions and `if` statements, use curly brackets `{}` to contain multiple statements between them. These brackets can be placed on separate lines or on the same line as the other code. For example:

```
if (allow == true) {
  alert(message) }
```

is the same as

```
if (allow == true)
{
  alert(message)
}
```

Variables

A *variable* is an alias that stores another piece of data — a text string, a number, a `true` or `false` value, or whatever. Variables are core to programming because they eliminate the need for having everything known while you are developing the script or application.

For example, suppose you wanted to capture the first name of the user and add it to a welcome message that appears when the user returns to your iPhone Web app. Without variables, you'd have to know the person's name beforehand — which is, of course, impossible — or you'd have to use a generic name for everyone.

By using a variable, you eliminate this problem. You can capture the name of the user and then store it in the code as a variable — I'll call it `firstName`. Then, anytime in my script I need to work with the user's name, I call `firstName` instead.

Declaring and Assigning Variables

When you want to use a variable, your first task is to declare it with the keyword `var` and then assign it a value. Here's how this two-step process can look:

```
var firstName;
firstName = "Jonah";
```

The `var` keyword instructs the JavaScript interpreter to treat the following word, in this case `first-Name`, as a variable. The following line then assigns the literal string `Jonah` to the `firstName` variable.

Although declaring and assigning a variable may be two separate actions, you'll usually want to combine them into one step. For example, the following code is functionally equivalent to the previous two lines of code:

```
var firstName = "Jonah";
```

Truth be told, you can actually assign a variable without even using the keyword `var`. For example, the following code also works:

```
firstName = "Jonah";
```

However, I strongly recommend regular use of `var` when you first define it to ensure your code is readable and manageable by both you and other developers.

Hanging Loose with Variable Types

You can assign a variety of data types to a variable. Following are a few examples:

```
// String value
var state = "CO";
// Boolean value
var showForm = true;
// Number value
var maxWidth = 300;
// HTML document "node"
var element = document.getElementById("container");
```

```
// Array item
var item = myArray[0];
```

If you are coming from an Objective C, Java, ActionScript, or other stronger-typed language, you are probably wondering where you declare the type of data that can be stored by the variable. Well, to put it simply — you don't have to. Because JavaScript is a loosely typed language, you don't need to declare the variable type.

Naming Variables

When you are naming variables, keep in mind the following general rules:

❑ A variable name must begin with a letter or underscore, not a number or other special character. For example:

```
// Valid
var helper;
var _helper;

// Invalid
var 2helper;
var ^helper;
```

❑ Although a number can't start off a variable name, you can include one elsewhere in the word, such as here:

```
var doc1;
var doc2;
```

❑ As with other aspects of JavaScript, variable names are case sensitive. Therefore, each of the following is a unique variable name:

```
var counter;
var Counter;
var COUNTER;
var counTer;
```

❑ A variable name can't contain spaces. For example:

```
// Valid
var streetAddress;
var street_address;

//Invalid
var street address;
```

❑ You can't use a reserved word (see Table 2-8) or the name of a JavaScript or Document Object Model object as a variable name. For example:

```
// Valid
var isDone;
var defaultValue;
var mainWindow;

// Invalid
```

```
var in;
var default;
var window;
```

Accessing a Variable

After you define a variable and assign it a value, each time you reference it in your code, the variable represents the value that was last assigned to it. In the following snippet of code, the `company` variable is assigned a string value and displayed in an alert box:

```
var company = "Cyberg Coffee";
alert( "Our company is proudly named " + company);
```

The alert box displays this:

```
Our company is proudly named Cyberg Coffee.
```

You can also change the value of a variable inside your source code. After all, they are called variables, aren't they? A typical example is a counter variable that has a value incremented during a looping process. Consider the following example:

```
for (var i = 1;i <= 5; i++)
{
   document.write( "The value of the variable is: " + i + ".<br/>" );
}
```

I'll explain what the `for` loop does later in the chapter. For now, just focus on the variable named `i`. The variable starts out with a value of 1 during the execution of the first pass of the code inside the `for` block. During the next pass, its value is incremented to 2. On the third pass, its value is incremented to 3. This same process continues until `i` is 5. In total, the `document.write()` method would be called five times and would output the following content:

```
The value of the variable is 1.
The value of the variable is 2.
The value of the variable is 3.
The value of the variable is 4.
The value of the variable is 5.
```

Scope

Variables can have either a limited or a wide scope. If you define a variable inside a function, that variable (called a *local variable*) is only available to the function and expires when the function is finished processing. For example, check out the `markup` variable in the code that follows:

```
function boldText(textToDisplay)
{
   var markup = "<strong>" + textToDisplay + "</strong>";
   return markup;
}
```

The variable is created when the `boldText()` function runs, but it no longer exists after the function ends.

However, if you define the variable at the top level of the scripting block — in other words, outside a function — that variable (known as a *global variable*) is accessible from anywhere in the document. What's more, the lifetime of a global variable persists until the document is closed.

Table 2-2 summarizes the differences between local and global variables.

Table 2-2: Local Versus Global Variables

Type	Scope	Lifetime
Local	Accessible only inside a function	Exists until the function ends
Global	Accessible from anywhere in the document	Exists until the document closes

Scope deals with the context of the variable within the scripting hierarchy. However, the location at which a variable is defined also matters. Simply put, a variable is accessible when it is declared. Therefore, you'll typically want to define global variables at or near the top of the script and local variables at or near the top of the function.

I've put together a sample script that demonstrates the scope of variables. First, I'll define a global variable at the top of a script:

```
<script language="javascript" type="text/javascript">
var msg = "I am a global variable";
var msgEnd = ", and you cannot stop me."
```

Next, I'll add an alert box to display the variable:

```
alert(msg + msgEnd);
```

When this code runs, the alert message text that appears will be as follows:

```
I am a global variable, and you cannot stop me.
```

Now check out what I am going to do. I am creating a function that assigns a local variable with the same name (msg) and then simply references msgEnd, the global variable I already defined. Here's the code:

```
function displayAltMessage()
{
  var msg = "Don't hassle me, I'm a local variable";
  alert(msg + msgEnd);
}
```

I call this function in my script with the following line:

```
displayAltMessage();
```

When this function is executed, the alert box displays the following text:

```
Don't hassle me, I'm a local variable, and you cannot stop me.
```

Notice that the local variable `msg` took precedence over the global variable `msg`. So within the function, the local `msg` variable "hides" the global `msg` variable, and any changes made to `msg` apply only to the local variable. However, when that function finishes, the local variable `msg` is destroyed, whereas the global variable `msg` remains.

Suppose that I reassign the value of the global variable `msg` later in my script and then display the alert box once again:

```
msg = "Reassigning now!";
alert(msg);
```

Yes, you've got it — the following text is displayed:

```
Reassigning now!
```

Notice that when I reassigned the variable, I didn't need to add the `var` keyword again.

Here's the full script:

```
<script type="text/javascript">
  // Global variable
  var msg = "I am a global variable";
  var msgEnd = ", and you cannot stop me."

  // Displays "I am a global variable, and you cannot stop me."
  alert(msg + msgEnd);
  function displayAltMessage()
  {
    // Local variable
    var msg = "Don't hassle me, I'm a local variable';
    // Displays "Don't hassle me, I'm a local variable, and you cannot stop me."
    alert(msg + msgEnd);
  }
  // Calls the function
  displayAltMessage();

  // Reassigning value to msg
  msg = "Reassigning now!";

  // Displays "Reassigning now!"
  alert(msg);
</script>
```

JavaScript Naming Conventions

Like Java and ActionScript, JavaScript uses CamelCase conventions for naming parts of the language. Variables, properties, and methods use what is known as lowerCamelCase. When using the *lowerCamelCase* convention, the name begins with a lowercase letter, but the first letter of each subsequent new word in a compound word is uppercase, with all other letters lowercase. You'll find that JavaScript and the DOM adhere to this convention. For example:

```
var divElements = document.getElementsByTagName("div");
```

In this example, `document` is the DOM object name for the HTML document and is all in lowercase. The method `getElementsByTagName()` is a compound word, so the first letter of each additional word in the method is capitalized. My variable `divElements` uses lowerCamelCase style to make it clear that the term combines two words.

Another type of coding convention that some JavaScript developers follow is what is known as *Hungarian notation*. Following this coding style, variables are prefixed with a letter that indicates the data type of the variable. Table 2-3 shows several examples of this notation.

In the end, the coding and naming convention you decide on is your personal or team preference. But no matter what you decide on, what ultimately matters is sticking with your conventions to create readable and manageable code.

Table 2-3: Using Hungarian Notation with Your Variables

Data/Object Type	Prefix	Example
String	S	sFirstName
Number	N	nRating
Boolean	B	bFlag
Array	A	aMembers
Object (DOM elements, user objects, and so on)	O	oCustomer
Date	D	dTransactionDate

Working with Constants

As you work with variables, they often change values during the course of a script. However, sometimes you want to work with a variable that refers to a value that doesn't change. These are called *constants*.

ActionScript, Java, and many other programming languages actually have a constant built into the language that is different from a variable. JavaScript does not. However, from a practical standpoint, you can define a variable and treat it as a constant in your code.

If you do, I recommend following standard conventions for naming constants:

- ❏ Use uppercase letters.
- ❏ Separate words with underscores.

For example:

```
var APP_TITLE = "iPhoneWebApp";
var VERSION_NUM = "2.01";
var MAX_WIDTH = 320;
```

You can then use these constant-looking variables just as you would a variable throughout your script.

Operators

An operator is a no-frills symbol used to assign values, compare values, manipulate expressions, or connect pieces of code.

The equal sign is perhaps the most common operator. It assigns a value to an expression. For example:

```
var i = 10;
```

Note that the operator that assigns a value (=) is different from the operator that compares whether two values are equal. The == operator is charged with comparing two items and returning `true` if conditions are true or `false` if not. For example:

```
alert(n == 100);
```

As you would expect, the + operator can add two number values together. However, you can also use it to combine two string literals or variables. So, check out this code:

```
var s = "Welcome ";
var t = " to the world of operators.";
alert(s + t);
```

The text `Welcome to the world of operators` would be displayed in a JavaScript alert box.

Tables 2-4 through 2-7 highlight the major operators. A few of the operators are pretty esoteric in real-world situations, but the ones you will typically use are shown in bold.

Table 2-4: Assignment Operators

Operator	Example	Description
=	x = y	The value of y is assigned to x
+=	x += y	Same as x=x+y
-=	x -= y	Same as x=x-y
*=	x *= y	Same as x=x*y
/=	x /= y	Same as x=x/y
%=	x %= y	Same as x=x%y (modulus, division remainder)

Table 2-5: Comparison Operators

Operator	Example	Description
==	x == y	x is equal to y
!=	x != y	x is not equal to y
===	x === y	Evaluates both for value and data type (e.g., if x = "5" and y = 5, then x==y is true, but x===y is false)
<	x < y	x is less than y
<=	x <= y	x is less than or equal to y
>	x > y	x is greater than y
>=	x >= y	x is greater than or equal to y
?:	x=(y < 5) ? -5 : y	If y is less than 5, then assign -5 to x; otherwise, assign y to x (known as the *conditional operator*)

Table 2-6: Logical Operators

Operator	Example	Description
&&	if (x > 3 && y=0)	Logical and
||	if (x>3 || y=0)	Logical or
!	if !(x=y)	Not

Table 2-7: Mathematical Operators

Operator	Example	Description
+	x + 2	Addition
-	x - 3	Subtraction
*	x * 2	Multiplication
/	x / 2	Division
%	x%2	Modulus (division remainder)
++	x++	Increment (same as x=x+1)
--	x--	Decrement (same as x=x-1)

Reserved Words

JavaScript has reserved words (see Table 2-8) that are set aside for use with the language. Avoid using these words when you name variables, functions, or objects.

Table 2-8: JavaScript 1.5 Reserved Words

abstract	boolean	break	byte	case
catch	Char	class	const	continue
debugger	Default	delete	do	double
else	Enum	export	extends	false
final	Finally	float	for	function
goto	If	implements	import	in
instanceof	Int	interface	long	native
new	Null	package	private	protected
public	Return	short	static	super
switch	Synchronized	this	throw	throws
transient	True	try	typeof	var
void	Volatile	while	with	

Basic Conditional Expressions

JavaScript has three conditional statements that you can use to evaluate code:

- ❑ `if`
- ❑ `if/else`
- ❑ `switch`

These three statements are explained in the following sections.

The if Statement

The `if` statement is used when you want to evaluate a variable and expression and then perform an action depending on the evaluation. The basic structure looks like this:

```
if (condition)
{
  // code to execute if condition is true
}
```

For example:

```
response = prompt("Enter the name of the customer.", "");
if (response == "Jerry Monroe")
{
    alert("Yes!");
}
```

A prompt box is displayed to the user. The typed response of the user is assigned to the `response` variable. The `if` statement then evaluates whether `response` is equal to "Jerry Monroe". If it is, the alert box is displayed. Otherwise, the `if` block is ignored.

I mentioned this in the operator section, but it bears repeating. The double equal signs are not a typo. Whereas a single equal sign (=) is used to assign a value to a variable, a double equal sign (==) compares one side of an expression to another.

If you are evaluating a Boolean variable, you can write the `if` statements in two ways:

```
if (isValid == true)
{
    alert("Yes, it is valid!");
}
```

Or, as a shortcut:

```
if (isValid)
{
    alert("Yes, it is valid!");
}
```

If you have a single line of code in the `if` block, you can actually forgo the curly brackets. So, this syntax works too:

```
if (isValid == true) alert("Yes, it is valid!");
```

You can also use `if` on a negative expression. Suppose, for example, that you want to perform an operation if a Boolean variable named `isValid` is `false`. The code would look something like this:

```
if (isValid == false)
{
    // do something to get validation
}
```

Notice that even though the `false` value is being used to evaluate the variable, the `if` block only executes if the expression (`isValid == false`) is `true`.

An alternative and usually preferred way to write this expression is to use the `not` operator (!):

```
if (!isValid)
{
    // do something to get validation
}
```

This expression says, in effect, if `isValid` is not true, then do something.

The if/else Statement

The `if/else` statement extends the `if` statement so that you can also execute a specific block of code if the expression evaluates to `true` and a separate block if it evaluates to `false`. The basic structure is this:

```
if (condition)
{
  // code to execute if condition is true
}
else
{
  // code to execute if condition is false
}
```

Here's an example of the `if/else` statement:

```
if (showName)
{
  document.write("My name is Sonny Madison.");
}
else
{
  document.write("I cannot disclose my name.");
}
```

Chaining if and if/else

You can also chain together `if` and `if/else` statements. Consider this scenario:

```
if (state = "MA")
{
  document.write("You live in Massachusetts.");
}
else if (state="CA")
{
  document.write("You live in California.");
}
else
{
  document.write("I have no idea where you live.");
}
```

The switch Statement

Besides chaining the `if` and `if/else` statements, you can use the `switch` statement to evaluate multiple values. Its basic structure is this:

```
switch (expression)
{
case label1 :
  // code to be executed if expression equals label1
  break;
```

```
case label2 :
  //code to be executed if expression equals label2
  break;
case label3 :
  //code to be executed if expression equals label3
  break;
default :
  // code to be executed if expression is different
  // from both label1, label2, and label3
}
```

The `switch` statement evaluates the expression to see if the result matches the first case value. If it does, the code inside the first `case` statement is executed. Note the `break` at the end of each of the `case` statements. This stops the `switch` statement from continuing to evaluate more of the `case` statements that follow. The `default` statement is executed if no matches are found.

Note that the `case` blocks do not use curly braces inside of them.

Here's an example that uses the `switch` statement to evaluate the current time of day using the built-in `Date()` object:

```
var d = new Date();
var hrs = d.getHours();

switch (hrs)
{
case 7 :
  document.write("Good morning.");
  break;
case 12 :
  document.write("It must be time for lunch.");
  break;
case 15 :
  document.write("Afternoon siesta. Zzzzz.");
  break;
case 23 :
  document.write("Bed time. Boo!");
  break;
default
  document.write("Check back later.");
}
```

The `hrs` variable is assigned the current hour using the `Date,getHours()` method. The `switch` statement evaluates `hrs`, examining whether this value matches the `case` statements in sequence. If it finds a match, the code inside the `case` is executed. Otherwise, the `default` statement is executed.

Loops

A common need that you will have in developing scripts in your iPhone Web apps is the ability to perform a given task several times or on a series of objects. Or maybe you need to execute code until a condition changes. JavaScript provides two programming constructs for these purposes: `for` and `while` blocks. A `for` loop cycles through a block of code *x* number of times. A `while` loop executes a block of code as long as a specific condition is `true`.

The for Loop

Along with the `if` statement, the `for` loop is likely the construct you'll use most often. The following snippet shows a `for` loop calling `document.write()` five times:

```
for (var i = 1; i <= 5; i++)
{
   document.write("Number:" i + "<br/>");
}
```

When run, the output looks like this:

```
Number 1
Number 2
Number 3
Number 4
Number 5
```

Going back to the code, notice the three parts of the `for` statement inside the parentheses:

❑   **Initialize variable:** The `var i = 1` initializes the variable used in the count. (The variable `i` is the typical name you'll see used for loop counters.)

❑   **Looping condition:** The `i <= 10` indicates the condition that is evaluated each time the loop is cycled through. The condition returns `true` as long as the `i` value is less than or equal to 10.

❑   **Update:** The `i++` statement is called after each pass of the loop completes. The `++` is the operator that increments the value of `i` by 1.

while and do/while Loops

A `while` loop also cycles through a block of code one or times. However, it continues as long as the expression evaluates to `true`. Here's the basic form:

```
while(expression)
{
   // statements
}
```

Here's an example:

```
var max = 100;
while(num > max)
{
   document.write("Current count: " + num);
   num++;
}
```

In this example, the code block loops as long as `num` is less than `100`. However, suppose that `num` has the value of `101`. If it does, the `while` code block is never executed.

The `do`/`while` loop is similar except that the condition evaluates at the end of the loop instead of at the start. Here's the basic structure:

```
do
{
    // statement
} while (expression)
```

In real-world practice, `while` loops are used more often than `do`/`while` ones.

Comments

As in HTML or other languages, you can add your own comments to your JavaScript code. Comments are useful when you go back into the code weeks or months later and need to understand again your code and the thought process behind it. Or, you might use comments to note specific places in the code that you need to rework later on.

You can add single-line or multiline comments.

Single-Line Comments

To add a single-line comment, add two slashes (`//`) to a line. Any text on the same line to the right of the slashes is considered a comment and is ignored by the interpreter. For example:

```
// Define two string variables
var s = "Welcome ";
var t = " to the world of operators.";
// Display concatenated values in an alert box
alert(s + t);
```

Multiline Comments

You can also define a comment that spans multiple lines. To do so, enclose the comments inside `/*` and `*/` marks. Here's what it looks like:

```
/*
 * Purpose: Converts Internet string date into Date object
 * RICH TO DO: Double-check that this works under all cases
 */
function getDateFromString(s)
{
    s = s.replace(/-/g, "/");
    s = s.replace("T", " ");
    s = s.replace("Z", " GMT-0000");
    return new Date(Date.parse(s));
}
```

Using JSDoc Comments

The Java world has a documentation generator called Javadoc, which is a tool that can generate HTML-based documentation from the comments a developer adds to source code. JSDoc is an open source tool, written in Perl, that provides similar functionality.

You can download JSDoc at `http://jsdoc.sourceforge.net`.

To enable a script for JSDoc, simply denote that comments you want included using a multiline comment, but with a slightly special syntax: open the comment with a slash and two asterisks (/**), and close the comment the normal way (*/). Here's an example:

```
/**
 * Converts Internet string date into Date object
 */
function getDateFromString(s)
{
  s = s.replace(/-/g, "/");
  s = s.replace("T", " ");
  s = s.replace("Z", " GMT-0000");
  return new Date(Date.parse(s));
}
```

You can also add several attributes that can be used to better describe the code. These attributes are prefixed with an @ sign. You can even add HTML tags for readability. Check this out:

```
/**
 * Converts Internet string date into Date object.
 * <p>
 * This function is called when the user selects the
 * Cyborg option in the left-side module.
 *
 * @param   s string containing the
 * @returns  the date as a Date object
 */
function getDateFromString(s)
{
  s = s.replace(/-/g, "/");
  s = s.replace("T", " ");
  s = s.replace("Z", " GMT-0000");
  return new Date(Date.parse(s));
}
```

For full details on available attributes, formatting commands, and syntax rules, check out `http://jsdoc.sourceforge.net` as well the Javadoc conventions page at `http://java.sun.com/j2se/javadoc/writingdoccomments`.

Even if you don't plan on using the JSDoc tool to create source code documentation, the JSDoc conventions are useful for creating well-documented code. Plus, you may find in the future that if you would like to use the JSDoc generator, you're all set.

Functions

A *function* is the basic building block or organizing unit of JavaScript. A foundation of a house is composed of many building blocks. In the same way, a JavaScript or Ajax script usually consists of a set of functions.

These building blocks may be easy to take for granted. But just as a beautifully decorated home means nothing apart from its foundation, a killer Ajax or JavaScript animation routine is pretty meaningless without these building blocks to rest upon.

Creating a Function

A *function* is a module of JavaScript statements that together perform a specific action. Here's a basic example:

```
function addAndDisplay(a, b)
{
   var total = a + b;
   alert(total);
}
```

To call the function, you do this:

```
addAndDisplay(3, 2);
```

There are a few things to note. The function is named `addAndDisplay()` and takes two arguments: a and b. An *argument* (also called a *parameter*) is a variable or literal value that you want to pass on to the function to process. In this case, the calling function sends to the 3 and 2 integer values the function to do something with (add them together).

Inside the curly brackets, the first line adds together the arguments and assigns the sum to the `total` variable. The second line displays the total in an alert box. When the alert box is closed, the function ends and returns control to the code that called it.

As you can see, a function

❑　Is called by another section of code to perform a process, return a result, or both

❑　Hides the details of its programming logic from other parts of the code

A function can return a value to the calling code. For example, check out the following function:

```
function getTotal(subtotal)
{
   var t = subtotal * .05;
   return t;
}
```

The `getTotal()` function receives `subtotal` as an argument, multiples it by the state tax percentage, and then returns the total value to the statement that called the function using the `return` command.

When `return` executes, the function stops executing at that line and returns to the calling function. For example:

```
function createProfile(el)
{
```

```
if (el == null)
{
  alert("Undefined DOM element. Cannot continue.");
  return;
}
else
{
  // do something really important to el
}
}
```

In this example, the function checks to see if the `el` parameter is `null`. If it is, the function ends processing after displaying an alert box. If not, the function continues.

Creating an Inline Function

JavaScript also supports inline functions, which enable you to define an unnamed function for use within a specific context. For example, suppose you want to display an alert message when the window loads. Using a normal function, you can structure your script as follows:

```
<script>
  function init()
  {
    alert("I am saying something important.");
  }

  window.onload = init;

</script>
```

The `onload` event is assigned the `init()` function as its event handler. However, you can also use an inline function and eliminate the specific function definition:

```
<script>
  window.onload = function() { alert("I am saying something important.") };
</script>
```

Another advantage is that you can reference variables within a specific scope without having to pass them to a function as arguments.

Overloading Functions

Function overloading is the concept of being able to pass varying amounts of arguments to a function. You can overload a function in JavaScript if you use the `arguments` property of a function. For example, suppose you wanted to create a function that added two numbers together and then returned the result. You could code it like this:

```
function add(a, b)
{
  return a + b;
}
```

However, your needs can change over time. Instead of two numbers, you might now need to add three or four or five numbers at a time. Instead of writing separate functions for each case, you could overload the `add()` function to do all these summing tasks:

```
function add()
{
  var t = 0;
  for (var i = 0, l = arguments.length; i < l; i++)
  {
    t += arguments[i];
  }
  return t;
}
```

The `add()` function creates a `for` loop to iterate through each of the supplied arguments. As it does so, it adds the current argument to the `t` variable. This total is then returned to the statement that called the function. Each of the following calls now works:

```
var t = add(2, 2);
var t1 = add(2, 4, 5);
var t2 = add(392, 1230, 3020202, 1, 2033);
```

You can also use the `typeof()` command to overload different data types inside the same function. For example, if I wanted to expand the functionality of `add` to "add" string values, I could modify the function as follows:

```
function add()
{
  var argType = typeof(arguments[0]);

  if (argType == "number")
  {
    var t = 0;
    for (var i = 0, l = arguments.length; i < l; i++)
    {
      t += arguments[i];
    }
    return t;
  }
  else if (argType == "string")
  {
    for (var i = 0, l = arguments.length; i < l; i++)
    {
      if (i == 0)
      {
        var s = arguments[i];
      }
      else
      {
        s += ", " + arguments[i];
```

```
      }
     return s;
    }
    else
      return "Unsupported data type";
  }
}
```

The `add()` function examines the type of the first parameter and assigns the result to `argType`. If the type is a number, the routine continues just like the earlier example. If the type is a string, however, the function combines the arguments into a single comma-delimited string. If the type is neither a number nor a string, an error message is returned.

Data Types

JavaScript allows you to work with specific data types:

- ❏ String
- ❏ Number
- ❏ Boolean
- ❏ Null
- ❏ Undefined
- ❏ Array
- ❏ Object

Strings, numbers, and Booleans are often known as *primitive data types,* `null` and `undefined` as *special data types*, and objects and arrays as *composite data types.*

Strings

A *string* is a collection of characters enclosed in single or double quotation marks. A string is also called a *string literal.* The following lines are examples of strings:

```
var wd = 'Kitchen sink';
var el = "<p>This is HTML.</p>";
var n = "3.14";
var char = 'z';
var special = "@#(*&^%$#@!~<>S"}{+=";
var blank = "";
```

As you can see, you can include most anything in a string — normal text, HTML tags, numeric values, special characters, and more.

Notice that you can have an empty string (`" "`), which is *not* the same as a `null` or `undefined` value.

Because a string can be enclosed in either single or double quotation marks, you can easily embed a quotation mark and apostrophe by using the opposite type pair to enclose the string literal. For example:

```
var str = '"I like living dangerously", said Jack. "I am enclosing a quote in this
string."';
var str2 = "I'm happy, aren't I?";
```

Alternatively, you can use the backslash character (\) to "escape" an apostrophe/quotation mark if the string is enclosed by the same type of quotation mark. Here are a couple cases:

```
var str2 = 'I\'m happy, aren\'t I';
var el = "<div id=\"container\"></div>";
```

The backslash character tells the interpreter to treat the following quote mark as part of the string rather than as the ending delimiter.

For many tasks, you can work with string values as string literals in the manner in which I described. However, if you are going to parse, manipulate, transform, or otherwise pick on a string, you can also treat it as a `String` object.

Numbers

The number data type represents any sort of numeric value — integer, real floating-point number, octal, and even hexadecimal numbers.

Table 2-9 lists the variety of number values that are valid within JavaScript.

Table 2-9: Examples of Numbers

Number	Description
32	Integer
.00223	Floating-point number
1.42e5	Floating-point number
0377	Octal value
0x32CF	A hexadecimal integer.

In addition to these normal number types, JavaScript contains a special value NaN, which stands for "not a number." As you might expect, if you perform an invalid math operation, you'll get NaN as the result.

Boolean Values

A Boolean type represents either a `true` or a `false` value. You typically use it to determine whether a condition in your code is `true` or `false`. For example, suppose you assign `false` to the `flagUser` variable:

```
var flagUser = false;
```

You can then evaluate the `flagUser` variable to determine its Boolean state in an `if` control structure. For example:

```
if (flagUser == true)
{
   alert("Hey User! You are hereby flagged.");
}
```

The `if` statement displays the alert box if `flagUser` evaluates to `true`.

Note that a Boolean value is *not* wrapped in quotation marks. So notice the difference:

```
// Boolean
var isABoolean = true;
var isAString = "true";
```

Null Values

When a variable or object contains no value, its value is expressed as `null`. (Not the literal string `"null"`, just `null`.) You can test whether a variable or object is `null` to determine if it was defined earlier in the code. For example:

```
if (myArray != null)
{
   doSomethingReallyImportant();
}
```

This code block evaluates the variable `myArray`. If its value is not equal to `null`, the condition evaluates to `true` and something really important happens.

A `null` value is not the same as an empty string, the number 0 or `NaN`, or the undefined data type, which I explain in the next section.

If you want to remove the contents of the variable, you can assign it a non-value by setting it to `null`:

```
if (loggedOut == true)
{
   userName = null;
}
```

So, in this example, if the `loggedOut` variable is `true`, the `userName` variable is reset.

Undefined Values

On first take, an undefined value sounds like just another name for `null`. But an `undefined` value is actually a different data type. An `undefined` value is returned by the interpreter when you access a variable that has been declared but never assigned a value. For example, the following snippet displays an alert box with `undefined` as the message.

```
var isSmiling;
alert (isSmiling);
```

You'll also get `undefined` if you try to call a property that does not exist. For example:

```
alert(document.theStreetThatHasNoName);
```

Arrays

So far, I've been talking about variables as individuals. An *array*, in contrast, is a collection of values. An array is grouped and indexed so that you can access a specific item stored in the group.

JavaScript stores arrays as `Array` objects, meaning that it has a variety of properties and methods. As you work with these members, you can perform numerous tasks.

Creating an Array

You can create a new array by assigning a variable to empty brackets:

```
var castMembers = [];
```

You can also create a new array by using `new Array()`. It goes like this:

```
var myArray = new Array();
```

However, using the `new` operator to create an array is considered old school and is not considered a good practice.

In some programming languages, you need to specify the number of items in an array at the time you define it. That's not so with JavaScript. It just increases the size of the array each time you add a new item to it.

Adding Items to an Array

To add items to the array, you simply assign a value to a specific index in the array. The *index* is designated by a value within brackets, such as the following:

```
var castMembers = [];
castMembers[0] = "Jerry";
castMembers[1] = "George";
castMembers[2] = "Elaine";
castMembers[3] = "Kramer";
```

Notice that, because arrays are zero-based, you start counting at 0 rather than 1.

You can also define the array items by passing values as parameters:

```
var primeNumbers = [1, 3, 5, 7, 11];
```

Storing Multiple Types in an Array

You can store most any kind of data in an array, including strings, numbers, Boolean values, `undefined` and `null` values, objects, and functions. You can even store an array within an array if you really want to. What's more, you can store various types of data within the same array. For example, consider the following array that contains a string, number, function, DOM object, and Boolean value:

```
var el = document.getDocumentById("easy_bake_oven");
var pieLove = ["pie", 3.14, function() { alert("Apple Pie") }, el, true];
```

Getting and Setting the Size of an Array

You can get the number of items in an array by accessing its `length` property. (Yes, an array is an object, so it has its own set of properties and methods.) For example:

```
var primeNumbers = [1, 3, 5, 7, 11];
alert(primeNumbers.length));
```

The alert box displays 5.

In most cases, you don't need to explicitly set the size of an array, because it dynamically resizes on its own when you add or remove items from it. However, when you need to specify an array's size, you can assign a value to the `length` property. For example, the following line increases the size of the array to 10:

```
primeNumbers.length = 10;
```

Note that no new data items are created when you do so.

If you specify a `length` that is smaller than the current size of the array, the items that fall outside of that `length` value get the axe.

Keep in mind that because array indices start at 0, the last item in the array will have an index value one fewer than the total number of elements in an array.

Accessing Items in an Array

You can access any value in the array according to its index. For example, to write the third item in a `states` array to the document, you can use the following code:

```
function showHomeState(idx)
{
    document.write("<p>Your home state is " + states[idx] + "</p>");
}
```

You can also iterate through an array using a `for` loop if you want to work with each item in the group:

```
var castMembers = [];
castMembers[0] = "Jerry";
castMembers[1] = "Jan";
castMembers[2] = "Elias";
castMembers[3] = "Bob";
document.writeln("<ul>");
```

```
for (var i = 0, l = castMembers.length; i < l; i++)
{
   document.writeln("<li>" + castMembers[i] + "</p>");
}
document.writeln("</ul>");
```

In this example, the `castMembers` array defines four string items. The `length` property determines the size of the array, so you know the number of times to loop through the `for` structure.

Notice that the `length` property is assigned to a variable rather than being included directly inside the `for` loop like this:

```
// Valid, but less efficient
for (var i = 0; i < castMembers.length; i++)
{
}
```

Although that is perfectly valid, it is not as efficient, particularly for large arrays. That's because the interpreter has to evaluate the `length` property on each pass rather than once. Therefore, it's a good habit to assign an array's length to a variable for loops.

The HTML output for this script is as follows:

```
<ul>
<li>Jerry</li>
<li>Jan</li>
<li>Elias</li>
<li>Bob</li>
</ul>
```

Passing by Reference or Value?

When a primitive variable (string, number, or Boolean) is passed into an array, the array creates a *copy* of the variable's value rather than storing a reference to that variable. In programming lingo, this is known as *passing by value*. Here's an example that demonstrates the difference:

```
var n = 100;
var b = true;
var s = "string";
ar = [n, b, s];

document.write("<p>[First pass]</p>");
for (var i = 0; i <= 2; i++)
   document.write("<p>"+ar[i]+"</p>");

n = 200;
b = false;
```

```
s = "new and improved string";

document.write("<p>[Second pass]</p>");
for (var i = 0; i <= 2; i++)
  document.write("<p>"+ar[i]+"</p>");
```

Notice that the variables n, b, and s are passed as parameters in the ar array. The for loop then outputs the values of the array items as HTML content to the document. The variables are then updated and given new assignments. However, when the same for loop is run again, the *original* values are output. Check out the HTML content that is output:

```
[First pass]
100
true
string
[Second pass]
100
true
string
```

However, when you pass an object or a function into an array, the array stores a reference to those advanced data types rather than a copy of it. As a result, any changes you make to the object or function outside the array are reflected when you access its corresponding item in the array. This is called *passing by reference.*

The Document Object Model

The Document Object Model (DOM) provides a scripting interface into an HTML document so that you can work with all its elements within a hierarchical structure. As a result, you can navigate through the document structure with JavaScript to access and manipulate anything inside of it.

The DOM is a standard that was originated by the W3C, the Web standards body. There are three DOM levels:

❑ **Level 1:** Level 1 contains the core functionality to be able to script the DOM for an HTML or XML document. Any modern browser provides support for the Level 1 version of the DOM.

❑ **Level 2:** Level 2 extends DOM support to include new core functionality, event listeners, CSS DOM, and traversal and range. Safari and other browsers generally provide strong support for aspects of DOM2.

❑ **Level 3:** Level 3 adds support for newer properties and methods, keyboard events, and XPath. Safari provides strong support for Level 3.

The DOM as a Tree

You'll find it helpful to think of the DOM as a hierarchical tree that contains all the parts of the HTML document — elements, attributes, text, comments, and so on. In DOM lingo, these individual pieces are known as *nodes*.

When working with HTML markup, the primary building block of your document is the element. However, when working with the DOM, the basic piece you will work with is the node. Remember that they are not identical. Although the DOM tree is dominated by element nodes, you need to be aware of several other types.

Just like a family tree in real life, each of the nodes is interconnected. A node can have a parent, a sibling, a child, and the occasional half-crazy "grandparent" that no one talks about. The document object is the family patriarch, serving as the container for all the nodes (the descendants) of the HTML file. The document object has no corresponding HTML element. The html, head, and body elements are contained by document.

One point I should stress is that a DOM hierarchy is *not* identical to the natural element hierarchy you find in an HTML document. Take, for example, the following plain vanilla HTML file:

```html
<html>
    <head>
        <title>Really Awesome Web Page</title>
    </head>
    <body>
        <div id="container">
            <h1>Pretty awesome heading</h1>
            <p id="para1">Pretty lame paragraph</p>
        </div>
    </body>
</html>
```

The levels of indentation show the natural element hierarchy in the document. Or, you can express the same hierarchy in a visual tree, as shown in Figure 2-4.

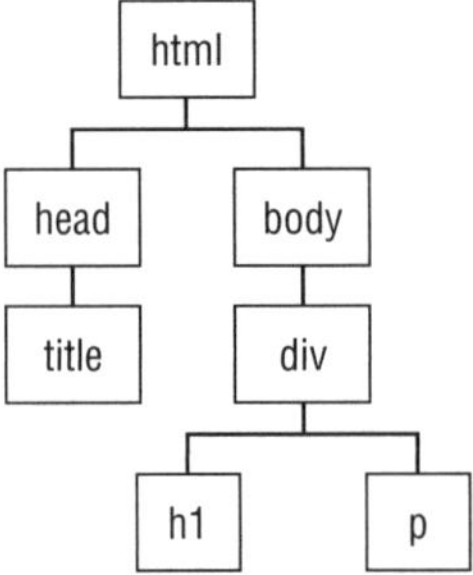

Figure 2-4: An element-based hierarchy is not the same as a DOM.

Every HTML element in the DOM has properties that correspond to the element's attributes. For example, an `img` has `src` and `alt` properties. However, DOM elements also have a set of common properties and methods.

In the DOM, elements are just one type of node. As a result, you have to express other nodes of the document, such as attributes and text content, in the tree. The text content inside an element is a child node of the parent element. So, for example, `Pretty lame paragraph` is considered a child node of the `p` element that contains it.

Attributes are also considered nodes, but they are a bit of a special case. They don't participate in the parent/child relationships that the other nodes do. Instead, they are accessed as properties of the element node that contains them. Figure 2-5 shows the same document in a DOM tree.

When working with the DOM in JavaScript, you typically work with the `document` object and DOM element objects. Each DOM element object has `childNodes` and `attributes` properties for accessing these nodes, as shown in Figure 2-6. And, as Figure 2-6 illustrates, each node has a `nodeType` property that can determine the type of node you are working with.

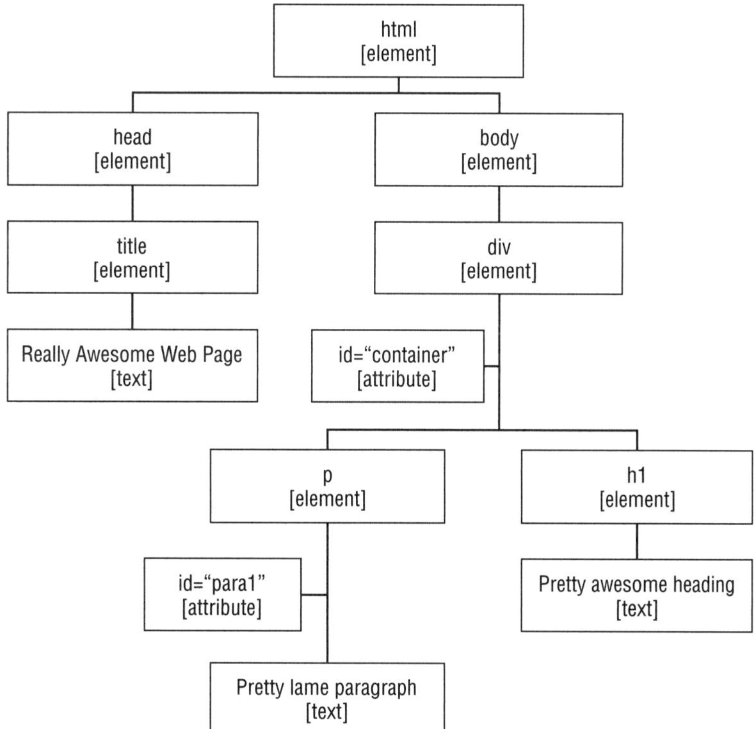

Figure 2-5: A DOM tree consists of many types of nodes.

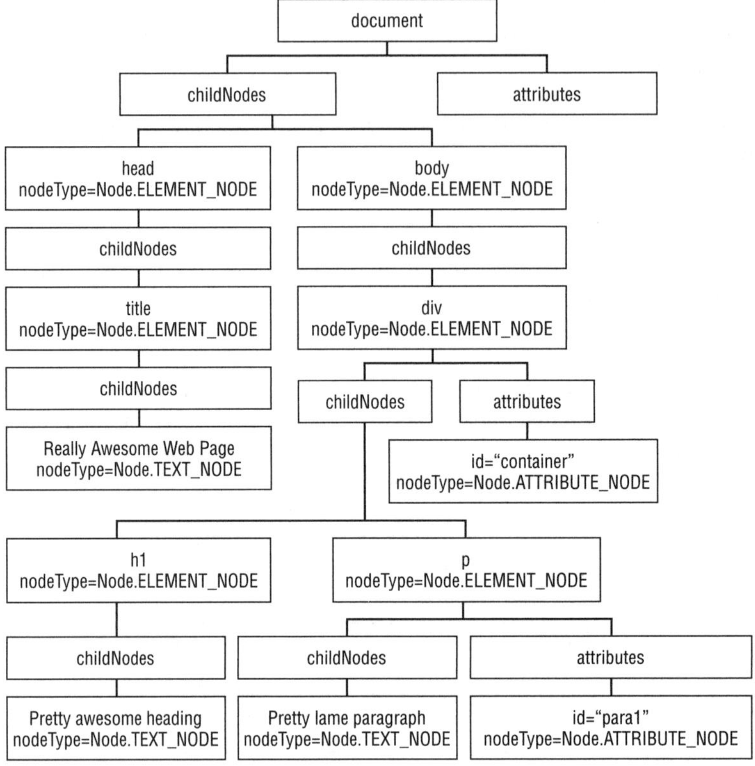

Figure 2-6: JavaScript's view of the DOM.

Understanding Node Types

Before you begin to access the DOM, it is important to take a moment to identify the node types that the DOM contains and that you'll work with. The `nodeType` property of each DOM node returns a constant from the `Node` object that represents the type of node. These include:

- ❏ `Node.ELEMENT_NODE`
- ❏ `Node.ATTRIBUTE_NODE`
- ❏ `Node.TEXT_NODE`
- ❏ `Node.CDATA_SECTION_NODE`
- ❏ `Node.ENTITY_REFERENCE_NODE`
- ❏ `Node.ENTITY_NODE`
- ❏ `Node.PROCESSING_INSTRUCTION_NODE`
- ❏ `Node.COMMENT_NODE`
- ❏ `Node.DOCUMENT_NODE`

❑   Node.DOCUMENT_TYPE_NODE

❑   Node.DOCUMENT_FRAGMENT_NODE

❑   Node.NOTATION_NODE

Each of the Node constants represents an unsigned short number. ELEMENT_NODE represents 1, ATTRIBUTE_NODE equals 2, and so on.

Accessing the DOM from JavaScript

You access the document object through a named object called, appropriately enough, document. Therefore, if you want to access a document object's property or method, you can do so directly. For example:

```
document.open();
```

You can access other elements and nodes in the DOM in a variety of ways, as shown in the following sections.

Accessing a Specific Element

When you know the id of an element, the most common way to access the element node is by using getElementById(). This method searches the DOM looking for the one node that has an id value matching the string you specify. For example, consider the following:

```
<html>
<body>
<div id="header">
<p>Rawhide!</p>
</div>
</body>
</html>
```

If you wanted to access the div, you could use the following line:

```
var d = document.getElementById("header");
```

The variable d now references the div id="header" element. You can now use the d variable to access its properties and perform methods on it. For example, to set its text color, you can use the following:

```
d.style.color = #cccccc;
```

Accessing a Set of Elements

In addition to accessing a specific element, you'll find occasions in which you'll want to access all element nodes of a given kind. You can also access a set of elements by using the document object's getElementsByTagName() method. This method takes an HTML element as its parameter:

```
var e  = document.getElementsByTagName(elementName);
```

This method returns all the elements with the specified tag name as a *NodeList,* or a collection of nodes ordered by their appearance in the document.

For example, to get all the p elements in a document and return it as a NodeList, you can type the following:

```
var cPara = document.getElementsByTagName("p");
```

After you have the collection of elements through either of these techniques, you can access a particular node in the NodeList by using its index number. For example, to access the first p element in the document, you can write this:

```
var p1 = cPara[0];
```

If you are accessing a specific p element, such as the first, you can combine both lines into a single call:

```
var p1 =  document.getElementByTagName("p")[0];
```

This returns the first p element and assigns it to the p1 variable.

More commonly, you can use it in a for loop to perform an action on each element:

```
for (var i = 0; i < cPara.length; i++)
{
    var para = cPara[i];
    para.style.color = #000000;
}
```

This code block iterates through each p element in the document and changes its text color to black. Notice that because the cPara variable is a nodeList, you can access its length property to determine the number of elements returned.

In addition, you can use getElementsByTagName() to return *all* element nodes inside the calling element by using * as the parameter:

```
var cElements = document.getElementsByTagName("*");
```

This call returns all the descending elements inside the document, regardless of the level of hierarchy they sit on.

Accessing Family Members

You can access nodes related to other nodes through more "familial ways" by using several node properties:

❑   parentNode returns the parent node of the current element.

❑   firstChild gets the first child of a node.

❑   lastChild gets the last child of a node.

- ❑ `previousSibling` gets the node just before the current one in the parent's NodeList.

- ❑ `nextSibling` gets the node just after the current one in the parent's NodeList.

- ❑ `childNodes` gets all the children of a node.

For example, you can use the `parentNode` property to get the parent node of a given element. For example, consider the following HTML snippet:

```
<div id="container">
<p id="p1"></p>
</div>
```

The following JavaScript returns a reference to the `div` element:

```
iParent = document.getElementById("p1").parentNode;
```

A parent node must be the `document`, the element node, or a document fragment.

In addition, suppose you wanted to return all the nodes that were direct children under an element. You could use the `childNodes` property instead to retrieve its children as a NodeList and store it in a variable. Here's what you would code:

```
var c = document.getElementById("container");
var cChildren = c.childNodes;
```

In this code, the element with an `id` of `container` is assigned to the `c` variable. Its children are then assigned to the `cChildren` variable and are ready for action.

You can also use the `hasChildNodes()` method to verify whether an element has child nodes. For example:

```
var c = document.getElementById("container");
if (c.hasChildNodes())
{
    var cChildren = c.childNodes;
}
```

Retrieving Attributes

Attributes are technically nodes in a DOM, but don't think of them as being equal to element nodes. You cannot, for example, retrieve a list of `id` attribute nodes through some sort of `GetAttributeByName()` method. In fact, except in rare occasions, there would be no practical reason for doing so because attributes are inextricably associated with an element. Not surprisingly, attributes are accessible through a DOM element by accessing its `attributes` property or by using one of two methods (hasAttributes() and hasAttribute()).

Accessing All Attributes

The object that returns from the `attributes` property is called `NamedNodeMap` object, which is similar to a NodeList except that its items are listed in arbitrary order, rather than a specific defined order.

If you want to retrieve a list of attributes for a given element, you might write something like the following:

```
var atts = document.getElementById("container").attributes;
```

You can then work with the attributes by accessing their `name` and `value` properties. For example:

```
var atts = document.getElementById("container").attributes;
var str = "";

for (var i = 0; i < atts.length; i++)
{
str += "Name:" + atts[i].name + " Value:" + atts[i].value + "<br/>";
}
```

This code block gets the attributes for an element and outputs the name and value for each attribute to the `str` variable.

You can check to see whether an element has attributes by using the `hasAttributes()` method:

```
var ctr = document.getElementById("container");
if (ctr.hasAttributes())
{
    var atts = ctr.attributes;
}
```

Accessing a Specific Attribute

You can also access a specific attribute using the element's `getAttribute()` method. Simply enter the name of the attribute as the parameter. Suppose, for example, you want to get the `href` attribute from a link:

```
var att = document.getElementById("book").getAttribute("href");
```

You can also check for the existence of an attribute with the `hasAttribute()` method. To verify whether an element has a `style` attribute, use the following code:

```
if (document.getElementById("contentMajor").hasAttribute("style"))
{
    var att = document.getElementById("contentMajor").getAttribute("style");
}
```

Manipulating the DOM

You can do more with the DOM than simply look at a document. In fact, you can also manipulate it — creating nodes, adding them to the DOM tree, editing nodes, and removing parts of a document you have no more use for.

When it comes to creating nodes, the document is the one in charge. The document creates all the nodes, and then you can insert it into the DOM tree at the appropriate location.

In the sections that follow, I'll show you how to add and remove various parts of a document to and from the DOM.

Creating an Element and Other Nodes

The document object has several methods for creating nodes. I'll focus on element nodes here.

To create an element, use the `document.createElement()` method. This method creates an instance of the element specified by the parameter. For example, to create a p element, use this:

```
var para = document.createElement("p");
```

The `createElement()` method creates the p element and returns the node, which is assigned to the `para` variable.

Note that the act of creating an element does not automatically add it to the DOM tree. The instance only exists in your script until you explicitly add it. It's in a sort of limbo land — a node without a country, if you will. Instead, you need to insert it into the document hierarchy to be a card-carrying node of the DOM.

Adding a Node to the DOM

DOM elements have two methods that you can use to add a node to the tree:

- ❑ `appendChild(node)` adds the node as a child at the end of the current element's `childNodes` list.

- ❑ `insertBefore(newNode, targetNode)` adds a new element as a child of the current node but inserts it just before the node specified by the `targetNode` parameter.

To show these in action, I'll be working with this HTML document:

```
<html>
  <head>
    <title>FavFilms</title>
  </head>
  <body>
    <div id="content">
      <h1 id="header1">Welcome to the World of Cinema</p>
      <p id="p1">Favorite films are shown below:</p>
      <ol id="movieList">
        <li id="film_30329">The Shawshank Redemption</li>
        <li id="film_30202">Casablanca</li>
        <li id="film_20216">Braveheart</li>
        <li id="film_28337">Groundhog Day</li>
      <ol>
    </div>
  </body>
</html>
```

Suppose that I wanted to add a new film onto the end of the `movieList`. I'd use the following code:

```
var fa = document.createElement("li");
document.getElementById("movieList").appendChild(f);
```

At the same time, I'd like to add a movie in the spot just before `Casablanca` by using `insertBefore()`:

```
var fb = document.createElement("li");
var casa = document.getElementById("film_30202");
document.getElementById("movieList").insertBefore(f, casa);
```

Although there is no `insertAfter()` method, you can perform the equivalent functionality by using `insertBefore()` in combination with the `nextSibling` property of the `targetNode`. So, supposing I wanted to add a node after `Braveheart`, I could use this:

```
var fb = document.createElement("li");
var br = document.getElementById("film_20216");
document.getElementById("movieList").insertBefore(f, br.nextSibling);
```

It's time to look at our results. The ordered list would now be updated to include the new element I added:

```
<ol id="movieList">
  <li id="film_30329">The Shawshank Redemption</li>
  <li/>
  <li id="film_30202">Casablanca</li>
  <li id="film_20216">Braveheart</li>
  <li/>
  <li id="film_28337">Groundhog Day</li>
  <li/>
<ol>
```

Creating Other Elements

The `document` object contains methods for creating the rest of the common node types: attribute, text, comment, and document fragment. Each of these methods returns a node that you can add to your document using `appendChild()` or `insertBefore()`.

An *attribute node* represents an element's attribute. To create an attribute, you use `createAttribute()`. For example:

```
var a = document.createAttribute("alt");
a.nodeValue = "Welcome to DOM and DOMMER.";
document.getElementById("logo").appendChild(a);
```

A *text node* is a container of text. When you are introduced to the DOM, it is probably natural to think of text being a property of an element. But text inside of an element is contained by a child text node. To create a text node, use `createTextNode()`. For example, the following code snippet creates a text node and adds it to the end of a p element with an `id` of `para`:

```
var t = document.createTextNode("I scream for ice cream.");
document.getElementById("para").appendChild(t);
```

A *comment node* represents an HTML comment. To create it, use `createComment()`:

```
var c = document.createComment("Insert new toolbar here.");
document.getElementById("toolbar_container").appendChild(c);
```

Finally, a *document fragment* is a temporary off-line document that you can use to create and modify part of a tree before you add the fragment to the actual DOM tree. When you're manipulating the DOM, it's much more efficient to work with a document fragment during the processing of a subtree (less overhead) than to always work directly with the actual DOM. To create a document fragment, use `createDocumentFragment()`. For example, the following script creates a document fragment, adds nodes to it, and appends it to the document:

```
var df = document.createDocumentFragment();

for (var i = 0; i < 30; i++)
{
  var e = document.createElement("p");
  var t = document.createTextNode("Line " + i);
  e.appendChild(t);
  df.appendChild(e);
}

document.getElementById("content").appendChild(df);
```

Now that you know how to create the other node types, I want to revisit the movie list example in the previous section. Instead of adding empty `li` elements, I want to add attribute and text nodes needed to fill out the ordered list appropriately.

First, I want to add a new film onto the end of the `movieList` using the following code:

```
var e; // Element
var a; // Attribute
var t; // Text
var nd; // Target/Parent

e = document.createElement("li");
a = document.createAttribute("id", "film_29299");
t = document.createTextNode("Babette's Feast");
e.appendChild(a);
e.appendChild(t);
document.getElementById("movieList").appendChild(e);

e = document.createElement("li");
a = document.createAttribute("id", "film_29323");
t = document.createTextNode("Amélie (Le Fabuleux destin d'Amélie Poulain)");
e.appendChild(a);
e.appendChild(t);
nd = document.getElementById("film_30202");
document.getElementById("movieList").insertBefore(e, nd);
e = document.createElement("li");
a = document.createAttribute("id", "film_30276");
t = document.createTextNode("Vertigo");
```

```
e.appendChild(a);
e.appendChild(t);
nd = document.getElementById("film_20216");
document.getElementById("movieList").insertBefore(f, nd.nextSibling);
```

The ordered list will now look the way I want it to:

```
<ol id="movieList">
  <li id="film_30329">The Shawshank Redemption</li>
  <li id="film_29323">Amélie (Le Fabuleux destin d'Amélie Poulain)</li>
  <li id="film_30202">Casablanca</li>
  <li id="film_20216">Braveheart</li>
  <li id="film_30276">Vertigo</li>
  <li id="film_28337">Groundhog Day</li>
  <li id="film_29299">Babette's Feast</li>
<ol>
```

Setting a Value to an Attribute

The `setAttribute()` method is a shortcut to creating an attribute and adding it to an element. This method sets the `attribute` value for an element, but if the attribute doesn't exist, the method creates it. It takes the attribute name and value as parameters:

```
setAttribute("name", "value")
```

For example, if you want to change the `href` of a link, you might use this:

```
document.getElementById("myblog").setAttribute("href",
"http://richwagnerwords.com");
```

Here's a little more detailed example of using `setAtttribute()` to clean up a document. Suppose you have an HTML document that has attributes with uppercase names. If you want to convert them to all lowercase, you can use the following script:

```
var all = document.getElementsByTagName("*");
for (var i = 0; i < all.length; i++)
{
  var e = all[i];

  for (var j = 0; j < all.length; j++)
  {
    e.attributes[j].setAttribute(all.attributes[j].nodeName.toLowerCase());
  }
}
```

The script looks for all the elements in a document and returns them as a `NodeList`. The first `for` loop iterates through all the elements and assigns the e variable to be the current element in the loop. The second `for` loop cycles through all the attributes of e and uses `setAttribute()` in combination with `toLowerCase()`.

Moving a Node

Although no `moveNode()` method is defined in JavaScript, `appendChild()` and `insertBefore()` do the same thing. In addition to adding new nodes you create, these two methods enable you to move a node from one location to another in the tree. For example, consider the following document:

```
<html>
  <head>
  </head>
  <body>
    <div id="header">
      <img id="logoImg" src="logo.png"/>
    </div>
    <div id="footer">
    </div>
    </body>
</html>
```

If you want to move the image from the header to the footer, you can write this:

```
var i = document.getElementById("logoImg");
document.getElementById("footer").appendChild(i);
```

When this code is executed, the `appendChild()` method moves the `logoImg` from its current location and adds it as a child to the `footer` div.

Cloning a Node

You can clone a node and add it elsewhere to the document tree by using the DOM element's `cloneNode()` method. This method copies the specified node and returns a new instance that is not part of the DOM. You can then add it wherever you want using the familiar `appendChild()` and `insertBefore()` methods.

This method takes a single Boolean parameter:

```
cloneNode(deepBoolean)
```

If the parameter is set to `true`, all subnodes of the current node are copied along with it.

When cloning nodes, you have to be careful. Every part of the node comes with using `cloneNode()`, including its `id`. Be sure to update the `id` before you add it to the document tree.

To demonstrate, I'll begin with the following document:

```
<html>
  <head>
  </head>
  <body>
    <div id="right_box">
```

```
        <p id="main">All content should begin with this sentence.</p>
      </div>
      <div id="left_box">
      </div>
      </body>
   </html>
```

If I want to copy the paragraph to the `left_box` div, I can enter the following:

```
var cn = document.getElementById("main_right").cloneNode(true);
cn.setAtttribute("id", "main_left");
document.getElementById("left_box").appendChild(cn);
```

Removing a Node from the DOM

To remove a node, call the `removeChild()` method of its parent, specifying it as the parameter. For example, to remove a paragraph with `id="p1"`, use the following code snippet:

```
var cp = document.getElementById("p1");
cp.parentNode.removeChild(cp);
```

In this code, the p variable is assigned the returning node of the specified paragraph element. Using `parentNode` to reference its parent, the `removeChild()` method is then called, specifying `cp` as the node to delete. The paragraph is removed from the DOM.

Or, suppose you want to remove all the children an element. To do so, you can loop through all the child nodes and delete them one at a time. Here's the code you can use:

```
while (bodyContent.childNodes[0])
{
   bodyContent.removeChild(bodyContent.childNodes[0]);
}
```

In this routine, the `while` loop checks to see if a child node exists for the current node. If it does, the child node is removed using `removeChild()`. This process repeats until there are no `childNodes[0]`.

Removing an Attribute

Suppose you have had it with an attribute and simply want to rid yourself of the pain and aggravation. Or maybe you just don't have a use for it anymore. If so, you can use the `removeAttribute()` method to delete an attribute from the DOM. The following rather draconian script removes all the `alt` attributes from the document — simply out of spite! Here's the code:

```
var kitAndKaboodle = document.getElementsByTagName("*");
for (var i = 0; i < kitAndKaboodle.length; i++)
{
   kitAndKaboodle[i].removeAttribute("alt");
}
```

Summary

In this chapter, I walked through some of the basics that will help you as you begin iPhone web app development. I began by showing you how to use and incorporate the new HTML 5 media elements into your web app to display rich media and serve as an alternative to using Flash or another unsupported technology. The chapter continued with an in-depth primer on the basics of JavaScript, which is a skill set you'll want to have as you begin to create functionality for your web apps. Finally, I closed out with a survey of the Document Object Model and how you can use it to find page data as well as manipulate it.

3

Building with Web App Frameworks

When you look up close at iPhone native and Web applications and compare their functionality, you will begin to see design patterns in how information is presented to a user. A list is presented this way. A page of menu options is presented that way. Some apps may break the rules, but by and large, applications follow common "look and feel" guidelines because they know users expect and rely on a consistent and intuitive way to perform a task.

To achieve that consistency and to avoid reinventing the wheel, developers often turn to application frameworks. In the generic sense, an application framework is a toolkit that you can use to implement a standard structure for an application for a given platform. Windows programmers utilize the Windows Software Development Kit (SDK) for creating apps for that desktop platform. So, too, developers writing native iPhone apps use the iPhone SDK.

In contrast, the Web development space often has been a virtual free-for-all, with each developer or shop coming up with its own toolkit to develop Web applications. Some programming languages may have application frameworks, but often these are designed for server-side issues, not for constructing the look and feel of an application over the Web.

As I've talked about already in this book, an iPhone Web app should be thought of from the ground up more along the lines of a native app than simply a Web site that users access from their iPhones. As such, the consistency and structure of an application framework can be extremely useful to the iPhone Web app developer.

In this chapter, I'll survey three popular freely downloadable iPhone Web app application frameworks that you should evaluate and consider adding to your toolkit. These include iWebKit, iUI, and UiUIKit. I cover each of them in slightly different detail. I spend the most amount of time with iWebKit as it offers considerable functionality and features. I introduce you to iUI, but explore it more in several chapters throughout this book. UiUIKit is less a framework than a set of HTML templates, so I provide an overview of its functionality.

Each of these frameworks gets to the same end—an iPhone UI—but does so using different techniques. Once you see each of these frameworks lined up against each other and compare their functionality differences, you can decide which one best suits your programming style and feature needs.

iWebKit

Home URL: `www.iwebkit.net`

License: GNU

iWebKit may not have been the first iPhone Web app framework available, but it has grown in popularity and maturity since its release. iWebKit consists of a CSS style sheet, a JavaScript script file, and a set of common UI images that you can utilize in your app. The `style.css` and `function.js` files come in compressed and uncompressed versions, enabling you to use the uncompressed version for testing and then pointing your app to the compressed version when you are ready to deploy.

iWebKit divides the UI into different `div` elements for the top bar, content area, footer, and other block regions. You then assign a specific `id` to the `div` to classify it as a specific type of UI element. Side-to-side menu navigation is structured using `ul` lists and `span` elements for headings. Within each `li` menu item, you define its link destination as well as other enhancements, such as an icon, arrow indicator, and special comments.

Figure 3-1 shows a page created using the iWebKit framework that takes advantage of many of its core features for a side-to-side navigation page. I will dive into the code that is used to create this page.

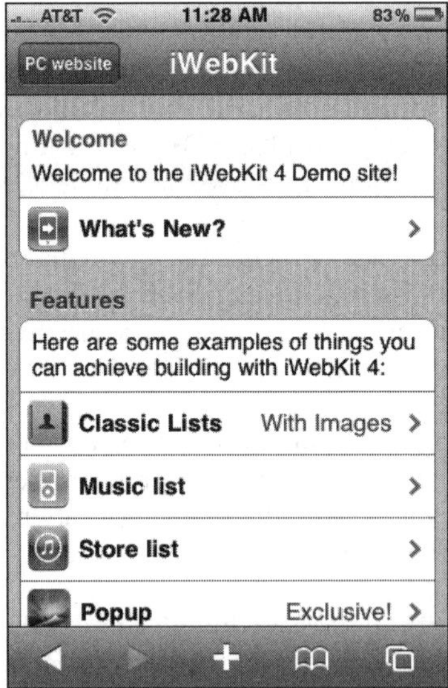

Figure 3-1: Page created using iWebKit's
side-to-side navigation framework

Enabling iWebKit

Along with standard iPhone Web app meta tags, every iWebKit-enabled page must include references to its CSS style sheet and JavaScript file in the document head:

```
<meta content="yes" name="apple-mobile-web-app-capable" />
<meta content="minimum-scale=1.0, width=device-width, maximum-scale=0.6667,
user-scalable=no" name="viewport" />
<link href="css/style.css" rel="stylesheet" type="text/css" />
<script src="javascript/functions.js" type="text/javascript"
language="JavaScript" ></script>
```

Core UI Elements

You'll incorporate several elements of the UI regardless of the application type or style. These are shown here.

The Top Bar

The top application bar of an iPhone Web app is typically used to display the title of your app/page and provide one or more "rabbit trail" buttons for navigating the app. To define the top bar, use the following `div` definition:

```
<div id="topbar"></div>
```

Inside of the top bar, you add a title:

```
<div id="topbar">
    <div id="title">iWebKit</div>
</div>
```

Then you add a navigation button:

```
<div id="topbar">
    <div id="title">iWebKit</div>
    <div id="leftbutton">
      <a href="http://iwebkit.net">PC website</a>
    </div>
</div>
```

The `id="leftbutton"` specifies a single left button to include on the top bar. You can also use `id="rightbutton"` to add a single button to the right side of the top bar.

Alternatively, you can add multiple navigation buttons to the top of a page by using either `<div id="leftnav"></div>` or `<div id="rightnav"></div>`. Multiple buttons enable you to achieve "rabbit trail" functionality, providing hierarchical links above or below the current page in the app. For example, notice the navigation for the page shown in Figure 3-2.

Figure 3-2: Navigation buttons on
the top bar

The following HTML was used to define the top bar shown in Figure 3-2:

```
<div id="topbar">
    <div id="leftnav">
        <a href="index.html"><img alt="home" src="images/home.png" /></a>
        <a href="list.html">Lists</a></div>
    <div id="rightnav">
        <a href="storelist.html">Store list</a>
    </div>
</div>
```

If you do plan to use multiple navigation buttons, you should avoid using a title tag, or you run a high risk of running out of real estate on the top bar.

The Content Region

The content region serves as a container for the main elements of a page. It is defined using the following tag:

```
<div id="content"></div>
```

Inside of the content tag, you can define several elements, depending on your needs.

The Content Box Container

The content region of the page shown in Figure 3-1 contains two menus, each of which is contained inside a white box container. The box container is added using a `ul` element:

```
<ul class="pageitem"></ul>
```

Inside this container, you can add different types of elements. These elements are `li` tags identified by their `class` name.

The Text Box

You can add labels or descriptive text using the following tag:

```
<li class="textbox"></li>
```

You can add a single line of text or multiple paragraphs inside the text box. For example:

```
<li class="textbox">
    <p>Description:</p>
    <p>This app does this and that...</p>
</li>
```

You can also add standard HTML content inside a text box. Figure 3-3 shows a text box containing a bulleted list. Here's the code for that:

```
<ul class="pageitem">
    <li class="textbox">
    <ul>
        <li>Popup now fully fullscreen compatible.</li>
        <li>Now the blue store list background loads only when it is needed,
        not always.</li>
        <li>Footer font styles now apply to the footer div not only the link
        in it.</li>
        <li>Automatic &quot;load 10 more&quot; in list and musiclist views.</li>
        <li>New wordpress icon.</li>
        <li>Various bug fixes and enhancements.</li>
    </ul>
    </li>
</ul>
```

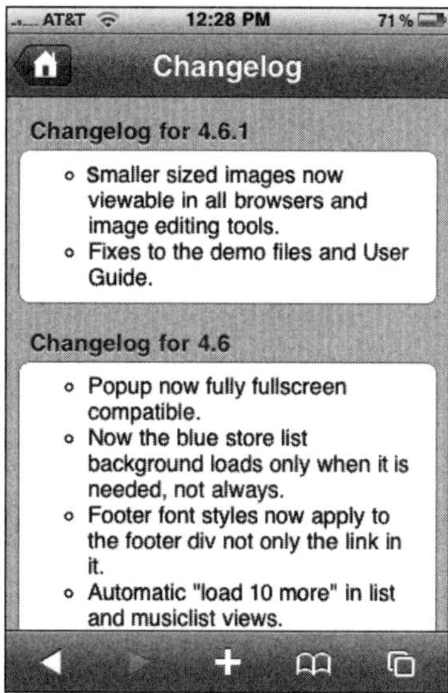

Figure 3-3: Bulleted list inside a text box

The Text Box Header

A blue header can be included inside a text box by surrounding the header content with the following:

```
<span class="header"></span>
```

That's what defines the Welcome header in Figure 3-1. As you can see, the heading is put in its own paragraph and then defined with the span:

```
<li class="textbox">
    <span class="header">Welcome</span>
    <p>Welcome to the iWebKit 4 Demo site!</p>
</li>
```

Menu Items

Menu items are at the heart of the navigation scheme of most iPhone apps. To define a text-only menu item inside a content box, use the following:

```
<li class="menu">
    <a href="goto.html">Menu item 1</a>
</li>
```

To add a bolded look to the menu item:

```
<li class="menu">
    <a href="goto.html">
        <span class="name">Menu item 1</span>
    </a>
</li>
```

To add an icon to the menu item, add an img tag inside the link:

```
<li class="menu">
    <a href="goto.html">
        <img alt="Description" src="thumbs/menuitem1.png" />
        <span class="name">Menu item 1</span>
    </a>
</li>
```

If you add an icon image, be sure it is a maximum of 32 pixels in height.

To add an arrow to the right-side:

```
<li class="menu">
    <a href="goto.html">
        <img alt="Description" src="thumbs/menuitem1.png" />
        <span class="name">Menu item 1</span>
        <span class="arrow"></span>
    </a>
</li>
```

The code that follows creates the What's New menu item shown in Figure 3-1:

```
<li class="menu">
    <a href="changelog.html">
        <img alt="changelog" src="thumbs/start.png" />
        <span class="name">What&#39;s New?</span>
        <span class="arrow"></span>
    </a>
</li>
```

You can also add special blue highlight text to insert a brief comment between the menu item and the right side arrow. The comment can call attention to the item ("Exclusive!") or can describe what users can expect to see if they select that option ("With Images"). Figure 3-4 shows a menu list in which three items add comments.

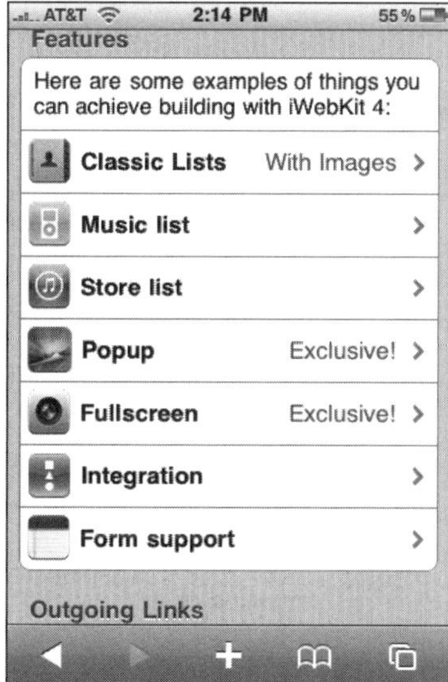

Figure 3-4: Comments can be added to menu items.

To add a comment, use a `<span class="comment"></span>` tag:

```
<li class="menu">
    <a href="list.html">
        <img alt="list" src="thumbs/contacts.png" />
        <span class="name">Classic Lists</span>
        <span class="comment">With Images</span>
        <span class="arrow"></span>
    </a>
</li>
```

The Gray Header

In the previous section, I showed you how to add a blue header inside a content box container. You can also add a gray header outside a content box. To do so, add the following as a direct child under the `<div id="content">` element:

```
<span class="graytitle">Features</span>
```

Refer to Figure 3-1 to see an example of the gray header in action.

The Footer

You can add a footer to the bottom of a page with the following declaration:

```
<div id="footer">
    <a href="http://iwebkit.net">Powered by iWebKit</a>
</div>
```

The footer `div` should be placed as a direct child of the `body` tag to get the gray-text-on-background look shown in Figure 3-5.

Figure 3-5: Footer added to the bottom of a page

Using the elements described in this section, you can create a basic iPhone side-to-side menu navigation page, such as the one shown in Figure 3-1.

Special Page Types

Besides the normal pages you can construct using the common UI elements described earlier, iWebKit features two special page types. However, when using them, you must follow their guidelines for the elements allowed in them.

A Simple List Page with Groupings

iWebKit allows you to create a simple list with optional groupings, such as the page shown in Figure 3-6. The advantage of the simple list is that the paging logic is built into the page type, allowing you to define the content and let iWebKit determine its pagination simply by adding an autolist element.

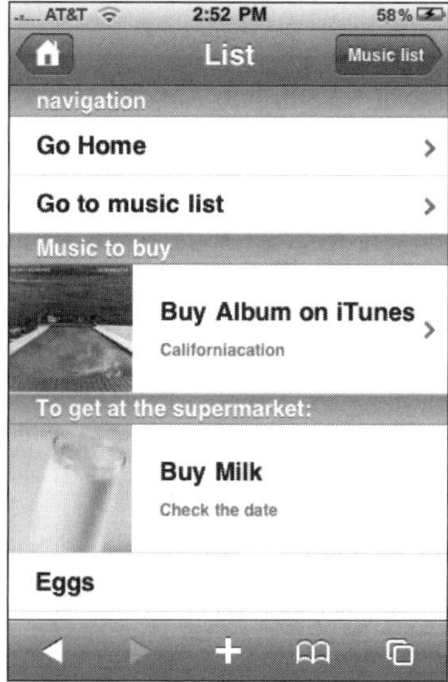

Figure 3-6: Simple list page type

To convert a normal page to a simple list page, add a `class="list"` attribute to the `body` tag:

```
<body class="list">
</body>
```

You should insert top bar, content, and footer `div`s into the body, just like any other iWebKit page. However, inside the content `div`, you'll want to define a special type of `ul` list:

```
<div id="content">
    <ul class="autolist">
    </ul>
</div>
```

All the content inside the `ul` should (naturally) be `li` items. The item itself can contain text and images and can contain links or static text.

The `autolist` class enables you to take advantage of iWebKit's automatic pagination, which I explain later in this section.

The Simple Text List Item

A normal list item text link looks essentially identical to the menu item described in the previous section:

```
<li>
    <a href="index.html">
        <span class="name">Go Home</span>
        <span class="arrow"></span>
    </a>
</li>
```

Because the `li` is contained inside the `ul` autolist, you don't need to identify a class for the `li`. A link, bolded item text, and right arrow are defined inside the item.

The Image List Item

Here's an example of a more complex list item that uses familiar elements (such as an image and comments). However, the simple list page formats them differently. Here's the code used for the Buy Album on iTunes item in Figure 3-6:

```
<li class="withimage">
    <a class="noeffect" href="http://itunes.apple.com/WebObjects/
    MZStore.woa/wa/viewAlbum?id=130244757">
        <img alt="test" src="pics/californication.jpg" />
        <span class="name">Buy Album on iTunes</span>
        <span class="comment">Californication</span>
        <span class="arrow"></span>
    </a>
</li>
```

The list item expands to fit the size of the image. Also note that the comments are listed below the item text and displayed in a smaller, grayed font.

The Nonlinked List Item

You may want to present information in a list structure that doesn't need hyperlinks. If so, just get rid of the `<span class="arrow"></span>` tag and then change the a link to simply be `<a class="noeffect"></a>`. For example:

```
<li>
    <a class="noeffect">
        <img alt="test" src="pics/milk.jpg" />
        <span class="name">Buy Milk</span>
        <span class="comment">Check the date</span>
    </a>
</li>
```

Pagination

iPhone UI conventions hold that if you are displaying a list of 10 or so items, it is fine to display them in a single one-page list. However, if you have 30 to 50 or more possible items, you'll want to break up the list into manageable chunks on the page. The logic to carry this out is not rocket science, but it sure is nice to have a framework do it for you. iWebKit does not disappoint; you can add pagination to your simple list by adding `class="autolist"` to the `ul` container and then the following line at the end of the list:

```
<li class="hidden autolisttext">
    <a class="noeffect" href="#">Load 10 more items...</a>
</li>
```

When you add this, iWebKit automatically splits up the list into multiple pages and adds the notification message you specify at the bottom (see Figure 3-7). When a user clicks the `Load 10 more items` link, iWebKit replaces the existing list contents with the next 10.

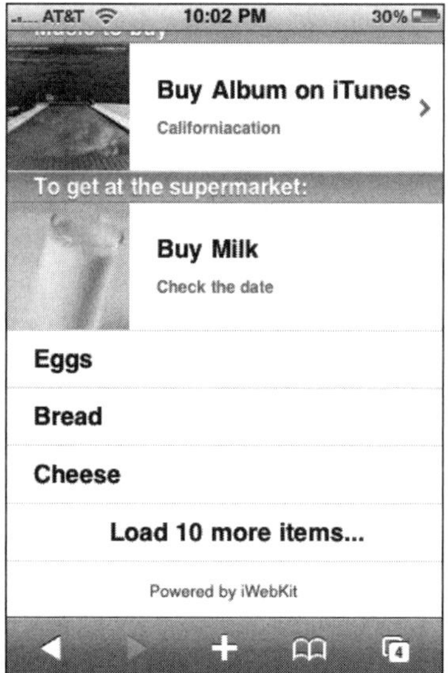

Figure 3-7: Splitting up a list of items

The Group Title

When you are dealing with a longer list of items, you may want to split up the list into multiple groups. To add a title to these subgroupings, use the following item:

```
<li class="title"> </li>
```

Figure 3-6 shows three group titles defined: `navigation`, `Music to buy`, and `To get at the supermarket`.

The Music List Page Type

A second special type offered by iWebKit is the Music list type, as shown in Figure 3-8. This page is structured as a two-column list, much like you could use to display a list of songs on an album. However, you can also use it for a variety of other purposes.

You define a music list page by adding the following class to the body tag:

```
<body class="musiclist"></body>
```

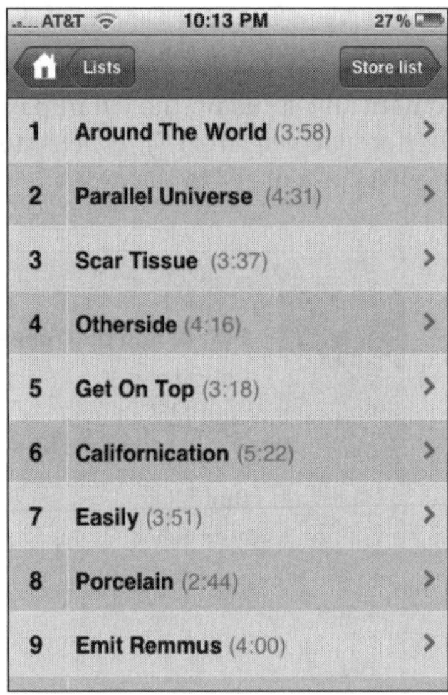

Figure 3-8: Music list page type

Much like the simple list, you can add a top bar, content region, and footer inside the body. Inside the content `div`, you define the following `ul` tag as a direct child of it:

```
<div id="content">
    <ul class="autolist">
    </ul>
</div>
```

The `class="autolist"` attribute specifies automated pagination.

Each item in a music list can have four fields of information: number, name, time, and link arrow. These are defined using spans, as shown here:

```
<li>
    <a class="noeffect" href="">
        <span class="number">1</span>
        <span class="name">Around The World</span>
        <span class="time">(3:58)</span>
```

```
            <span class="arrow"></span>
        </a>
    </li>
```

Form Elements

iWebKit provides support for styling form elements to emulate the look of iPhone form controls. To specify a form element, enclose the following within a special `li` tag:

```
<ul class="pageitem">
    <li class="form"></li>
</ul>
```

Inside the `<li class="form"></li>` tag, you can then specify different form elements.

To define a large text field, use an `<input type="text">` tag:

```
<li class="form"><input placeholder="Name" type="text" /></li>
```

The `placeholder` attribute value is added as grayed text in the text box. When the user clicks in it, the placeholder text disappears.

To define a large password field, use an `<input type="password">` tag.

```
<li class="form"><input placeholder="Password" type="password" /></li>
```

Like any password field, Safari substitutes dots for the characters that the user displays.

Figure 3-9 shows the large text and password fields.

Figure 3-9: Large text and password fields

To define a narrow text field with a label before it, use the `<span class="narrow">` tag combined with the `<span class="name">` tag.

```
<li class="form">
    <span class="narrow">
        <span class="name">First Name</span>
        <input type="text" />
    </span>
</li>
```

Figure 3-10 shows the results.

Figure 3-10: Narrow labeled field

You can create several other controls in a similar fashion. To define a text area, add a `textarea` inside of a `<li class=textbox">` tag:

```
<li class="textbox">
    <textarea name="TextArea">Add your text here</textarea>
</li>
```

To define a check box:

```
<li class="form">
    <span class="check">
        <span class="name">Label</span>
        <input name="remember" type="checkbox" />
    </span>
</li>
```

Users can't slide these yes/no controls, but pressing them toggles between the two values.

You can define a set of radio buttons by enclosing a set of radio inputs inside a `ul` container and assigning the same `name` attribute value for the `input` tags. When you do so, a user can select a mutually exclusive option from the set. Here's how you can define a three-option radio button set. (Note that the common name for the input name values is `music`):

```
<ul class="pageitem">
    <li class="form">
        <span class="choice">
            <span class="name">option1</span>
            <input name="music" type="radio" value="other" />
        </span>
    </li>
    <li class="form">
        <span class="choice">
            <span class="name">option2</span>
            <input name="music" type="radio" value="Metal" />
        </span>
    </li>
    <li class="form">
        <span class="choice">
          <span class="name">option3</span>
          <input name="music" type="radio" value="Metal2" />
        </span>
    </li>
</ul>
```

Figure 3-11 shows the check box and radio option groups.

Figure 3-11: Defining check box and radio options

To create a select drop-down group, simply define an HTML `select` element and stuff it inside an `li` tag:

```html
<li class="form">
    <select name="select">
        <option value="1">option1</option>
        <option value="2">option2</option>
        <option value="3">option3</option>
        <option value="4">option4</option>
    </select>
</li>
```

As you can see by now, there is a clear pattern for several form elements: define an HTML form element and place it inside an `<li class="form">` tag. Here are additional examples:

Submit button:

```html
<li class="form">
    <input name="Submit" type="button" value="submit" />
</li>
```

Reset button:

```html
<li class="form">
    <input name="Reset" type="button" value="reset" />
</li>
```

Button:

```html
<li class="form">
    <button name="button1">OK</button>
</li>
```

Popup Dialogs

One of the real innovative capabilities of iWebKit is that it enables you to create popup dialog boxes that emulate native iPhone dialogs (also known as *action sheets*) and are more sophisticated than the typical JavaScript alert box.

You define a popup dialog by using the `<div class="popup"></div>` tag. You should add this tag outside of the `<div id="content">` as a direct child of the `body`, but place all popup definitions before the `<div id="footer">`. Here's a sample popup dialog:

```html
<div id="popup1" class="popup">
    <div id="frame" class="confirm_screen">
    <span>popup example</span>
        <a href="index.html">
            <span class="gray">Home</span>
        </a>
        <a href="storelist.html">
            <span class="red">Previous feature</span>
        </a>
        <a class="noeffect" onclick="iWebkit.closepopup(event)">
```

```
            <span class="black">Cancel</span>
        </a>
    </div>
</div>
```

The `<span>` defines dialog box text, whereas the remaining a link definitions define different button styles: gray, red, and black. Figure 3-12 shows the dialog displayed in Safari.

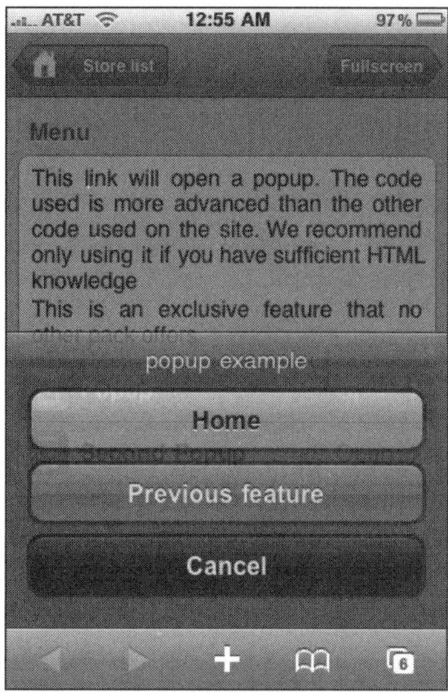

Figure 3-12: iWebKit's popup dialog

To call this dialog, you need to add an `onclick` handler to a link tag:

```
<a class="noeffect" onclick="iWebkit.popup('id')">
```

When the link is clicked, the `popup()` method of the `iWebKit` object is called. The `id` parameter specifies the `id` of the `div` that defines the popup.

Table 3-1 provides a summary of all the major UI elements that can be defined using iWebKit.

Table 3-1: iWebKit UI Elements

Component	HTML Code
Top bar	`<div id="topbar"></div>`
Top bar title	`<div id="title"></div>`
Top bar button (left)	`<div id="leftbutton"><a></a></div>`
Top bar button (right)	`<div id="rightbutton"><a></a></div>`

Continued

Table 3-1: iWebKit UI Elements *(continued)*

Component	HTML Code

Top bar navbar (left)	`<div id="leftnav">`
	`<a></a>`
	`<a></a>`
	`. . .`
	`</div>`

Top bar navbar (right)	`<div id="rightnav">`
	`<a></a>`
	`<a></a>`
	`. . .`
	`</div>`

Content region	`<div id="content"></div>`
Content box	`<ul class="pageitem"></ul>`
Text box	`<li class="textbox"></li>`
Text box header	`<span class="header"></span>`

Menu item	`<li class="menu">`
	`<a>Menu item</a>`
	`</li>`

Menu item with icon	`<li class="menu">`
	`<a>`
	`<img />`
	`Menu item`
	`</a>`
	`</li>`

Menu item with right arrow	`<li class="menu">`
	`<a>`
	`Menu item`
	`<span class="arrow"></span>`
	`</a>`
	`</li>`

Menu item with highlight text	`<li class="menu">`
	`<a>`
	`Menu item`
	`<span class="comment"></span>`
	`</a>`
	`</li>`

Component	HTML Code
Gray content header	`<span class="graytitle"></span>`
Footer	`<div id="footer"></div>`
Simple list page type	`<body class="list">` `</body>`
Simple list group heading	`<li class="title"></li>`

Simple list item (with link)	`<li>`
	`<a>`
	`Menu item`
	`<span class="arrow"></span>`
	`</a>`
	`</li>`

Simple list item (with image and comments)	`<li class="withimage">`
	`<a>`
	`<img />`
	`<span class="comment">Comment</span>`
	`<span class="arrow"></span>`
	`</a>`
	`</li>`

Music list (2-column) page type	`<body class="musiclist"></body>`

Large text form field	`<li class="form">`
	`<input placeholder="Name" type="text" />`
	`</li>`

Large password form field	`<li class="form">`
	`<input placeholder="Password" type="password" />`
	`</li>`

Narrow text field with labels	`<li class="form">`
	`<span class="narrow">`
	`<span class="name">Label</span>`
	`<input type="text" />`
	`</span>`
	`</li>`

Continued

Table 3-1: iWebKit UI Elements *(continued)*

Component	HTML Code

Check box	```<li class="form">```
	```<span class="check">```
	```<span class="name">Label</span>```
	```<input name="remember" type="checkbox" />```
	```</span>```
	```</li>```

Radio button group	```<ul class="pageitem">```
	```<li class="form">```
	```<span class="choice">```
	```<span class="name">Label</span>```
	```<input name="radiogroup1" type="radio"```
	```value="myvalue" />```
	```</span>```
	```</li>```
	```...```
	```</ul>```

Select list	```<li class="form">```
	```<select name="select">```
	```<option value="1">option1</option>```
	```...```
	```</select>```
	```</li>```

Text area	```<li class="textbox">```
	```<textarea></textarea>```
	```</li>```

Submit button	```<li class="form">```
	```<input name="Submit" type="button" value="submit"```
	```/>```
	```</li>```

Reset button	```<li class="form">```
	```<input name="Reset" type="button" value="submit" />```
	```</li>```

Button	```<li class="form">```
	```<button name="button1">OK</button>```
	```</li>```

Component	HTML Code
Popup dialog	``` <div id=" " class="popup"> <div id="frame" class="confirm_screen"> </div> </div> ```

iUI

Home URL: code.google.com/p/iui/

License: New BSD

iUI is the granddaddy of iPhone Web app frameworks. Introduced by developer Joe Hewitt soon after the original iPhone launch, it was the framework of choice for all early adopters. Joe eventually stopped working personally on the iUI framework, but he opened it as an open source Google code project.

iUI offers less special features than either iWebKit or UiUIKit, opting to focus on clean side-to-side navigation pages, destination pages, and core UI controls. And, as you'll see in the walkthrough, it provides automated control over parts of the app.

Figure 3-13 shows a demo app built using iUI. As I begin to explore the iUI code that drives the app, you'll begin to see the similarities and differences that iUI has compared to iWebKit.

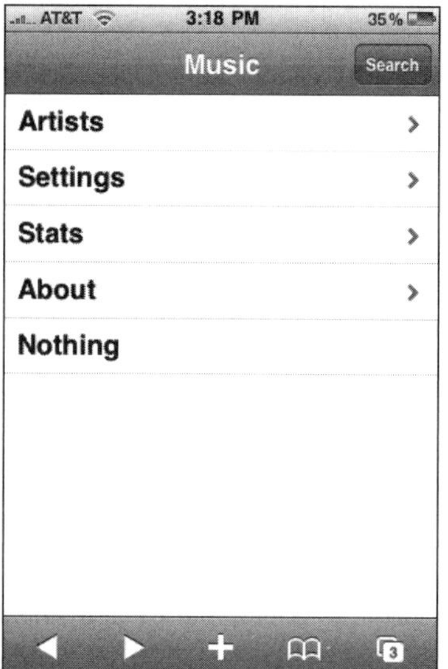

Figure 3-13: Web app built using the iUI framework

Structuring the App

One of the unique aspects of iUI is that screens or pages displayed in an iUI app are HTML fragments; for example, a menu page is a `ul` element, and a destination page is a `div`. You can house these HTML fragments inside a single HTML file, or you can separate them into separate files. However, if you separate them into separate files, be sure you don't reference normal HTML files (with an `html`, `head`, and `body`). Instead, the file should contain just the HTML element that represents the page.

Enabling iUI

To enable the iUI framework for a page in your Web app, you'll want to reference its CSS style sheet and JavaScript file in the document head:

```
<meta content="yes" name="apple-mobile-web-app-capable" />
<meta name="viewport" content="width=device-width; initial-scale=1.0;
maximum-scale=1.0; user-scalable=0;" />
<link href="iui/iui.css" rel="stylesheet" type="text/css" />
<script src="iui/iui.js" type="text/javascript" language="JavaScript"></script>
```

Once you have these declarations added, you are ready to begin defining the different parts of the interface.

The Top Bar

The top bar of the page is called a toolbar in iUI lingo. You can add one by defining a `div` and assigning it the `toolbar` class:

```
<div class="toolbar">
</div>
```

Each "screen" of the Web app defined inside the HTML file utilizes the same toolbar.

A top bar title can be added to the `div` by defining an `h1` tag with a `pageTitle` id:

```
<h1 id="pageTitle"></h1>
```

Don't place content in the page title, because iUI adds that automatically for you for each screen.

For a left-aligned Back button, add the following:

```
<a id="backButton" class="button" href="#"></a>
```

iUI keeps track of the pagination order, so leave the `href` defined as `href="#"`.

You can add a right-side action button to the top bar by defining an `a` link with the `class` button:

```
<a class="button" href="#searchForm">Search</a>
```

This button, which appears on each screen, points to a `#searchForm` section that I'll define shortly.

Combined, the application-wide top bar looks like this:

```
<div class="toolbar">
    <h1 id="pageTitle"></h1>
    <a id="backButton" class="button" href="#"></a>
    <a class="button" href="#searchForm">Search</a>
</div>
```

The Main Screen

Each screen (or page) is defined as a `ul` list, with each menu item as separate list items. Here is the code used for the main screen of the app shown in Figure 3-13:

```
<ul id="home" title="Music" selected="true">
    <li><a href="#artists">Artists</a></li>
    <li><a href="#settings">Settings</a></li>
    <li><a href="stats.php">Stats</a></li>
    <li><a href="http://code.google.com/p/iui/" target="_self">About</a></li>
    <li>Nothing</li>
</ul>
```

The `title` attribute of the `ul` is placed in the top bar title. Each list item defines a link that can point to another `ul` list, another HTML file, or an external URL.

To link to another `ul` section in the same file, add a # and its `id`. For example, the first list item links to the `<ul id="artists"></ul>` element (which is defined later). If you reference a `ul` in a separate HTML file, link to the file itself.

iUI automatically adds right-aligned arrows to each menu item.

A Simple List Page

A listing page (much like directory listing inside the native Contacts app) can be defined much like the previous screen:

```
<ul id="artists" title="Artists">
    <li class="group">B</li>
    <li><a href="#TheBeatles">The Beatles</a></li>
    <li><a href="#BelleSebastian">Belle &amp; Sebastian</a></li>
    <li class="group">C</li>
    <li><a href="#CrowdedHouse">Crowded House</a></li>
    <li class="group">J</li>
    <li><a href="#JennyLewis">Jenny Lewis</a></li>
    <li><a href="#JohnMayer">John Mayer</a></li>
    <li class="group">Z</li>
    <li><a href="#Zero7">Zero 7</a></li>
</ul>
```

You can add group headings by assigning a list item with the `group` class. Four group headings are defined in the preceding list. Figure 3-14 shows the list displayed inside Safari on iPhone.

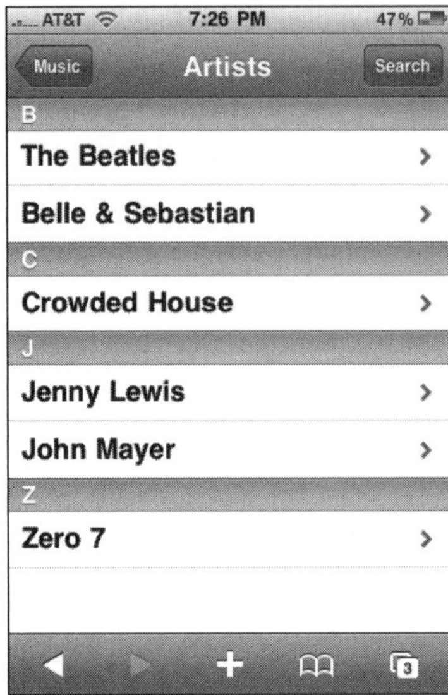

Figure 3-14: Grouped menu listing

Notice that iUI updates the text and link of the Back button to point to the page that linked to it.

Form Controls

iUI provides a basic level of form controls, such as the ones shown in Figure 3-15.

Figure 3-15: Form controls of iUI

Consider how this page is constructed using the iUI framework. The destination "panel" page is defined using the following definition:

```
<div id="settings" title="Settings" class="panel">
</div>
```

As with the menu page, the `title` attribute is used for the top bar title. The `panel` class gives the page its striped background.

Inside a destination page, you can define a heading using an ordinary `h2` tag:

```
<h2>Playback</h2>
```

A rounded corner box is defined with a `fieldset` tag. For example, the next code defines a fieldset. Inside of it, it defines two `div`s that define toggle buttons:

```
<fieldset>
    <div class="row">
        <label>Repeat</label>
        <div class="toggle" onclick="">
            <span class="thumb"></span>
            <span class="toggleOn">ON</span>
            <span class="toggleOff">OFF</span>
        </div>
    </div>
    <div class="row">
        <label>Shuffle</label>
        <div class="toggle" onclick="" toggled="true">
            <span class="thumb"></span>
            <span class="toggleOn">ON</span>
            <span class="toggleOff">OFF</span>
        </div>
    </div>
</fieldset>
```

The `<div class="row">` defines a row inside a `fieldset` that can contain different types of elements or controls. In this example, it contains a label and a toggle button.

As you can see from the code, the toggle button is not defined with a single element, but through a `div` container and a set of `span`s that are stitched together by iUI to create the control.

The second `fieldset` follows a similar pattern, except standard text input elements are defined:

```
<h2>User</h2>
<fieldset>
    <div class="row">
        <label>Name</label>
        <input type="text" name="userName" value="johnappleseed" />
    </div>
    <div class="row">
        <label>Password</label>
        <input type="password" name="password" value="delicious" />
```

```
        </div>
        <div class="row">
            <label>Confirm</label>
            <input type="password" name="password" value="delicious" />
        </div>
    </fieldset>
```

Table 3-2 summarizes the main iUI elements.

Table 3-2: Common iUI UI Elements

Component	HTML Code
Top bar	`<div class="toolbar"></div>`
Top bar title	`<h1 id="pageTitle"></h1>`
Top bar back button	`<a id="backButton" class="button" href="#"></a>`
Top bar action button (right-aligned)	`<a class="button" href="#linkTo">Label</a>`
Menu-based page	`<ul id="pageId" title="pageTitle"></ul>`
Menu item	`<li><a href=""></a></li>`
Group item	`<li class="group"></li>`
Destination panel page	`<div id="id" title="pageTitle" class="panel"></div>`
Destination page heading	`<h2>Heading</h2>`
Rounded corner white box	`<fieldset></fieldset>`
Row inside box container	`<div class="row"></div>`

iUI is used in several examples throughout this book. Refer to these for more information on how to implement iUI in your apps.

UiUIKit

Home URL: `code.google.com/p/iphone-universal/`

License: GNU General Public License v3

UiUIKit (Universal iPhone UI Kit) is more like a starter pack for creating iPhone Web apps, a set of customizable graphics and templates that you can then extend for your own purposes. Unlike iWebKit and iUI, it does not include a JavaScript library for performing certain tasks, such as pagination, or automatically performing some of the styling. However, it does support a truly impressive set of CSS styles to emulate a vast array of iPhone UI elements that go well beyond normal side-to-side navigation and basic destination pages.

Figure 3-16 shows the main page of a sample UiUIKit app. Notice that its style, although similar, deviates slightly from iUI's more literal approach to Apple guidelines. In this section, I will show you the basic styles of UiUIKit and survey the special pages and elements you can re-create using this framework.

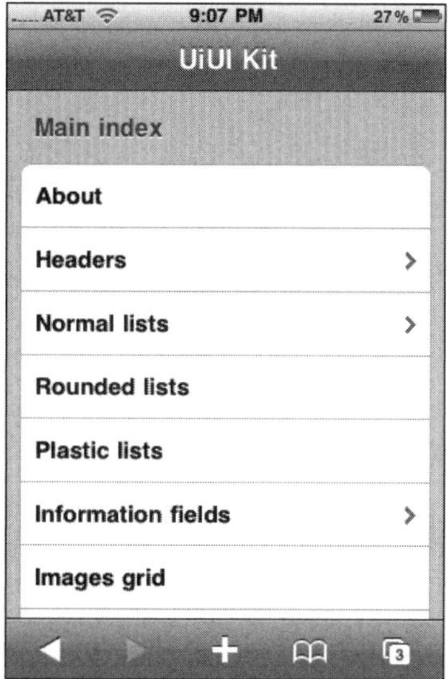

Figure 3-16: Form controls of iUI

Enabling UiUIKit

To use the UiUIKit framework in your Web app, you just need to reference its CSS style sheet file in the document head:

```
<meta content="yes" name="apple-mobile-web-app-capable" />
<meta name="viewport" content="width=device-width; initial-scale=1.0;
maximum-scale=1.0; user-scalable=0;" />
<link rel="stylesheet" href="stylesheets/iphone.css" />
```

Once you have this declaration added, you are ready to begin defining the different parts of the interface.

The Top Bar

The top bar of an app is defined using a `div` with a `header` id:

```
<div id="header">
</div>
```

To add a page title, use an h1 tag:

```
<h1>UiUI Kit</h1>
```

A Back button is defined by adding a `class` and `id` attributes:

```
<a href="index.html" class="nav" id="backButton">Index</a>
```

Right-side action buttons are defined by specifying nav `Action` as the class name:

```
<a href="search.html" class="nav Action">Search</a>
```

The Side-to-Side Menu List Page

You should define side-to-side menu navigation pages by adding an `id` of normal to the `body` tag:

```
<body id="normal"></body>
```

As with iUI, the actual menu is defined using a `ul` list along with child list items. For example, here's the HTML code used for the menu shown in Figure 3-16:

```
<ul>
    <li><a href="about.html">About</a></li>
    <li class="arrow"><a href="headers.html">Headers</a></li>
    <li class="arrow"><a href="normal-lists.html">Normal lists</a></li>
    <li><a href="rounded-lists.html">Rounded lists</a></li>
    <li><a href="plastic-lists.html">Plastic lists</a></li>
    <li class="arrow"><a href="info-lists.html">Information fields</a></li>
    <li><a href="images-list.html">Images grid</a></li>
    <li><a href="chat.html">Chat</a></li>
    <li><a href="forms.html">Forms</a></li>
    <li><a href="button-panels.html">Buttons Panel</a></li>

</ul>
```

As you can see, to add an arrow to the right of a menu item, you would add the `arrow` class to the `li`. Other than that, it looks like a standard HTML list.

To define highlight text on the right, add a `<small>` tag. For example:

```
<li><small>new!</small> <a href="normal-icon.html">Icon list</a></li>
```

You can also add extended ASCII characters inside the small tag to display symbols like stars and checkmarks.

UiUIKit does not stop there. It goes well beyond these standard side-by-side features and offers many more options.

To create a contact list, with group headings and name emphasis, you can combine heading tags and `ul` lists for each subsection of the menu. Here's a snippet of HTML that makes up the contact list page shown in Figure 3-17:

```html
<h4>E</h4>

<ul>
<li><a href="index.html">Elea <em>Peliche</em></a></li>
<li><a href="index.html">Elizabeth <em>Nogales</em></a></li>
<li><small>Es mi hermano</small> <a href="index.html">Emiliano
<em>Martín Lafuente</em></a></li>
<li><a href="index.html">Enrique <em>Dans</em></a></li>
<li><a href="index.html">Ernesto <em>González Aro…</em></a></li>
</ul>

<h4>F</h4>

<ul>
<li><a href="index.html">Fernández <em>Cols Aleix</em></a></li>
<li><a href="index.html">Fernández <em>Dominguez Alex</em></a></li>
<li><a href="index.html">Flavia <em>Olmedo</em></a></li>
<li><a href="index.html">Francesco <em>Esplugas</em></a></li>
<li><a href="index.html">Francisco <em>Perez Garzón</em></a></li>
</ul>
```

The only difference in code is the addition of the h4 group headings and the em tags to change the formatting for last names.

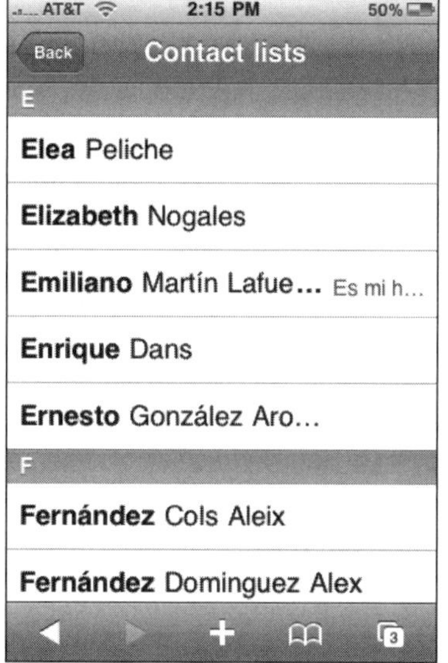

Figure 3-17: Contact list using UiUIKit

To add icons to the menu items, just add an `img` tag with a `class="ico"` attribute. For example, here's an icon menu:

```
<h1>Normal list with icons</h1>
<ul>
<li><img src="images/list-icon-1.png" class="ico" /> Example one</li>
<li><a href="normal-contact.html"><img src="images/list-icon-2.png"
class="ico" /> Example two</a></li>
<li><small>new!</small> <a href="normal-contact.html">
<img src="images/list-icon-3.png" class="ico" /> Example with label</a></li>
<li class="arrow"><a href="normal-content.html">
<img src="images/list-icon-4.png" class="ico" /> Example with arrow</a></li>
<li><a href="normal-metal.html"><img src="images/list-icon-5.png" class="ico" />
Metal list</a></li>
</ul>
```

Figure 3-18 displays this menu inside of Safari on iPhone.

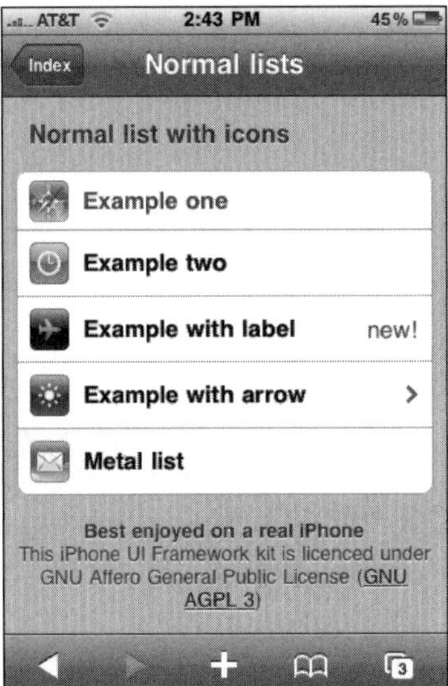

Figure 3-18: Icon menu items

The Destination Page

I did not mention it at the same time, but refer to Figure 3-18 for a moment. Notice that the menu is contained within a box, rather than being spread out from side to side. The reason for this is that the `body` tag did not have a `id="normal"` attribute added to it. Therefore, when a page has a plain `body` tag, the page is considered a destination page and is given the gray striped background.

You can define various elements on a destination page.

To define a heading on a destination page, use h1 as a direct child of the body:

```
<h1>Standard with content</h1>
```

The Plastic Page

UiUIKit includes a page type that neither iWebKit or iUi does—something it calls a plastic page (see Figure 3-19). A plastic page loosely resembles the look of the iPhone App Store—changing both the page background and menu appearances. What's more, you can define image-based banners and mini banners.

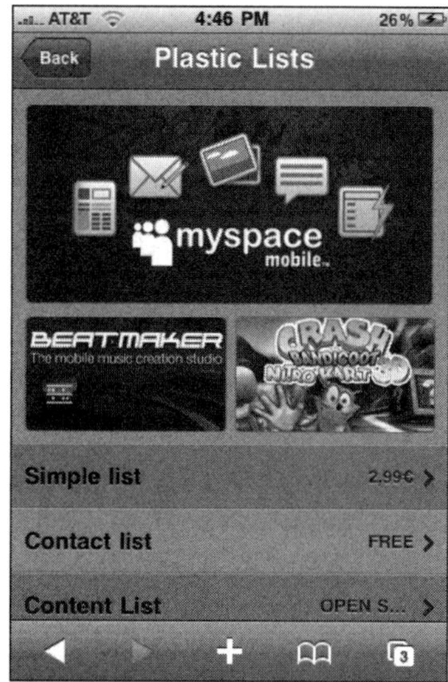

Figure 3-19: UiUIKit's plastic page

To define a plastic page, you need to give the body tag a plastic id:

```
<body id="plastic"></body>
```

Large and mini banners can be defined, but you need to write some style definitions yourself. Here's the code for the banners shown in Figure 3-19:

```
<ul class="bigbanner">
<li class="one"><a href="index.html">Use this space as title tooltip</a></li>
</ul>

<ul class="minibanner">
<li class="one"><a href="index.html">Use this space as title tooltip</a></li>
<li class="two"><a href="index.html">Use this space as title tooltip</a></li>
</ul>
```

However, to reference the specific images, you need to add the following CSS code to the document head:

```
<style type="text/css" media="screen">
ul.minibanner li.one { background: url(images/banner-1.png) no-repeat; }
ul.minibanner li.two { background: url(images/banner-2.png) no-repeat; }
ul.bigbanner li.one { background: url(images/banner-3.png) no-repeat; }
</style>
```

Plastic lists are structured the same way as the previous side-to-side menu lists created earlier. Because it is on a plastic page, UiUIKit transforms its look for you automatically:

```
<ul>
<li class="arrow"><small>2,99?</small> <a href="normal-simple.html">
Simple list</a></li>
<li class="arrow"><small>Free</small> <a href="normal-contact.html">
Contact list</a></li>
<li class="arrow"><small>Open Source</small> <a href="normal-content.html">
Content List</a></li>
<li class="arrow"><small>35?</small> <a href="normal-metal.html">Metal list</a>
</li>
<li class="arrow"><a href="normal-metal.html">Metal list</a></li>
<li class="arrow"><a href="normal-content.html">Content List</a></li>
<li><a href="/flash/player_net9_20080425.swf?link=594&cate=2">Download</a></li>
</ul>
```

UiUIKit offers some additional styles for specific pages as well. See Figures 3-20, 3-21, and 3-22.

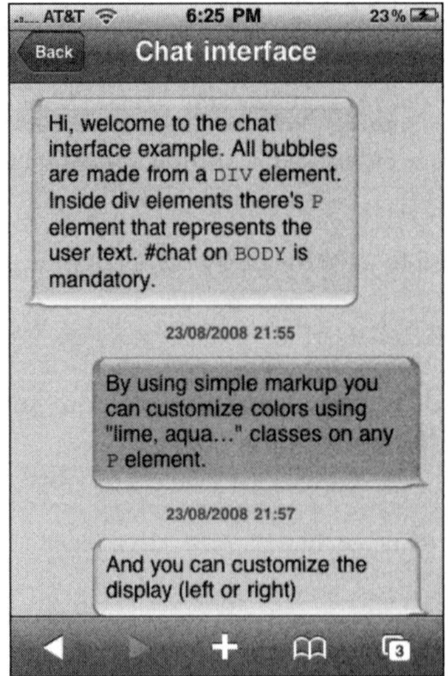

Figure 3-20: Chat balloon page

Figure 3-21: Image grid page

Figure 3-22: Contact page

Table 3-3 summarizes the most commonly used components of UiUIKit.

Table 3-3: Common UiUIKit UI Controls

Component	HTML Code
Top bar	`<div class="header"></div>`
Top bar title	`<h1></h1>`
Top bar back button	`<a href="url" class="nav" id="backButton">Label</a>`
Top bar action button	`<a href="url" class="nav Action">Label</a>`
Side-to-side navigation page	`<body id="normal"></body>`
Destination page	`<body></body>`
Plastic page	`<body id="plastic"></body>`

Navigation menu	`<ul>`
	`<li></li>`
	`...`
	`</ul>`

Menu item	`<li><a></a></li>`
Menu item (with arrow)	`<li class="arrow"><a></a></li>`

Menu item (with highlight text)	`<li>`
	`    <small>Comment</small>`
	`    <a href="url"></a>`
	`</li>`

Menu item (with icons)	`<li>`
	`    <a href="url">`
	`     <img src="" class="ico" />`
	`     Label`
	`    </a>`
	`</li>`

Group heading (side-to-side menus)	`<h4>Label</h4>`
Heading (destination page)	`<h1>Label</h1>`

Summary

In this chapter, I walked you through three popular iPhone web app frameworks that you can use with your iPhone web applications. iWebKit and iUI are full-fledged frameworks that can serve as the foundation for your web app. iWebKit is more feature-rich, while iUI aims to stick with standard side-to-side navigation and destination pages. UiUIKit is less of a framework than a set of styling templates that you can use as a basis for starting out with your app and as a way to avoid recreating the wheel on standard iPhone look and feel. Be sure to spend some time with each of these frameworks to determine which one is best for your particular needs.

Designing a Usable
and Navigable UI

User interface design has been evolutionary rather than revolutionary over the past decade. Most would argue that Mac OS X and Windows 7 both have much more refined UIs than their predecessors. As true as that may be, their changes improve upon existing ideas rather than offer groundbreaking new ways of interacting with the computer. Web design is no different. All the innovations that have transpired — such as Ajax and HTML 5 — have revolutionized the structure and composition of a Web site, but not how users interact with it. Moreover, mobile and handheld devices offered a variety of new platforms to design for, but these were either lightweight versions of a desktop OS or a simplistic character-based menu.

Enter the iPhone.

The iPhone interface is not a traditional desktop interface, although it has a codebase closely based on Mac OS X. It is not a traditional mobile interface either, despite that it is obviously a mobile device. You build Web apps using Web technologies, but the iPhone interface is not a normal Web application interface either.

Because the underlying guts of iPhone Web applications are based on tried-and-true Web technologies, many will be tempted to come to the iPhone platform and naturally want to do the same things they've always done — except customizing it for the new device. That's why the biggest mindset change for developers is to grasp that they are creating iPhone Web apps, not Web applications that run on iPhone. The difference is significant. In many ways, iPhone Web applications are far more like Mac or Windows desktop applications — users have a certain look and feel and core functionality that they will expect to see in it.

On the Web, users expect every interface design to be a one-off. Navigation, controls, and other functionality are usually unique to each site. However, when working on a platform — be it Windows, Mac OS X, or iPhone — the expectation is much different. Users anticipate a common way to perform tasks — from application to application. Operating systems provide application program interfaces (APIs) for applications to call to display a common graphical user interface (GUI).

Because iPhone web apps do not have such a concept, it is up to the application developer to implement such consistency.

This chapter provides the high-level details and specifications you need to consider when designing a UI for iPhone. The rest of the book builds upon this by diving into the actual code needed to implement these user interfaces.

The iPhone Viewport

A *viewport* is a rectangular area of screen space within which an application is displayed. Traditional Windows and Mac desktop applications are contained inside their own windows. Web apps are displayed inside a browser window. A user can manipulate what is seen inside the viewport by resizing the window, scrolling its contents, and in many cases, changing the zoom level. The actual size of the viewport depends entirely on the user, although the average size for a desktop browser is roughly 1000 • 700 pixels.

The entire iPhone display is 320 • 480 pixels in portrait mode and 480 • 320 in landscape. However, application developers don't have access to all that real estate. Instead, the viewport in which an iPhone developer is free to work with is a smaller rectangle: 320 • 416 in portrait mode without the URL bar displayed (320 • 356 with the URL bar shown), and 480 • 268 in landscape mode (480 • 208 with the URL bar). Figures 4-1 and 4-2 show the dimensions of the iPhone viewport in both orientations.

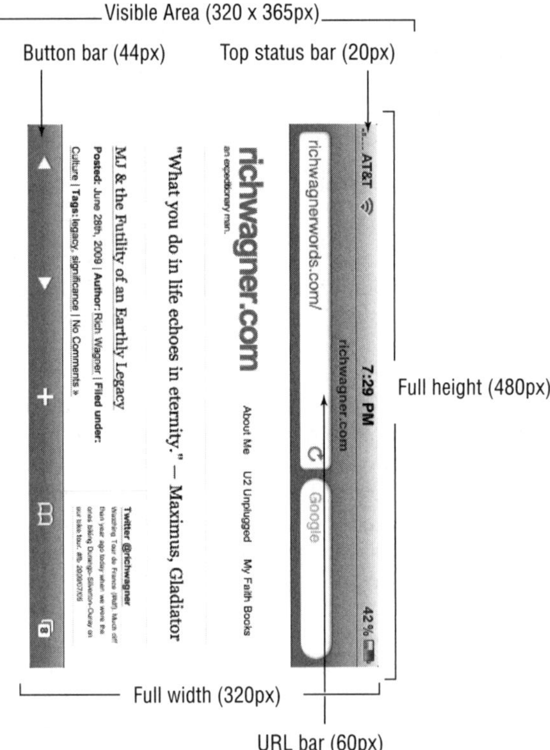

Figure 4-1: Portrait viewport

Figure 4-2: Landscape viewport

Users can scroll around the viewport with their fingers. However, they cannot resize it. To use desktop lingo, an iPhone application is always "maximized" and takes up the full available space.

If the on-screen keyboard is displayed, the visibility of the viewport is further restricted with the keyboard overlay, as shown in Figures 4-3 and 4-4.

Figure 4-3: Forms in Portrait viewport

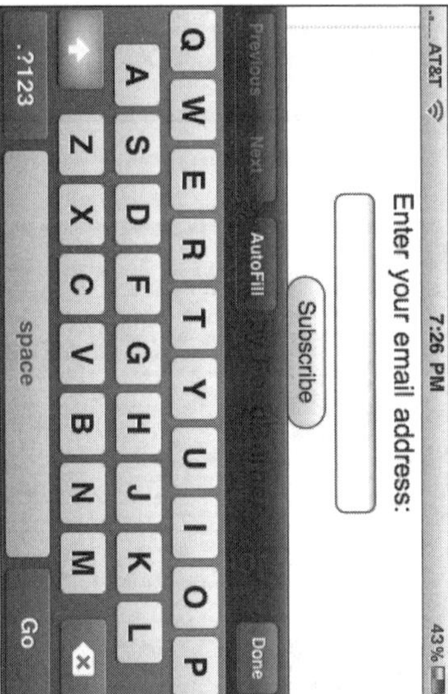

Figure 4-4: Landscape viewport

Because users have a much smaller viewport than they are used to working with on their desktop, the iPhone viewport has a scale property that can be manipulated. When Mobile Safari loads a Web page, it automatically defines the page width as 980 pixels, a common size for most fixed-width pages. It then scales the page to fit inside the 320 or 480 pixel-width viewport. Although 980 pixels may be acceptable for browsing a scaled-down version of ESPN.com or CNN.com, an iPhone application will almost certainly want to avoid this type of scaling by customizing the viewport meta tag.

The viewport meta tag sets the width and scale of the viewport. If you are creating an iPhone Web app, you'll want to set the viewport to be the exact width of the device — 320px in portrait mode and 480px in landscape mode. To make things easier, Safari on iPhone supports constants so you can avoid the specific numeric values. Therefore, to set the viewport for a normal iPhone Web app, you should add the following meta tag to the head of your HTML document:

```
<meta name="viewport" content="width=device-width; initial-scale=1.0;
maximum-scale=1.0; user-scalable=no;"/>
```

The width=device-width attribute sets the width to a fixed size and ensures that the viewport is not resized when the user rotates to landscape mode. The initial-scale=1.0 sets the scale at 1.0, whereas the maximum-scale and user-scalable settings disable user zooming of the Web page.

Exploring Native iPhone Applications

Before you begin designing your iPhone Web application, a valuable exercise is exploring the native iPhone applications on the iPhone or from the App Store. As you do so, you can consider how other designers handle a small viewport as well as how to design an intuitive interface for touch screen input.

However, to fully appreciate the design decisions that went into these applications, you need to understand the differences in the way in which users use iPhone applications compared to their desktop counterparts. After all, consider the types of applications that you will find installed on your desktop computer. An overly simplistic categorization is as follows:

❑ **Task-based applications:** The typical desktop application, whether it is on Mac, Windows, or Linux, is designed to solve a particular problem or perform a specific task. These applications (such as Word, Excel, PowerPoint, Photoshop, or iCal) tend to act upon one file or a few files at a time. The UI for these applications is often quite similar, including a top-level menu, a toolbar, common dialogs for open/save, a main destination window, and side panels.

❑ **Aggregators:** The second category of desktop application is aggregators — those applications that manage considerable amounts of data. You tend to work with many pieces of data at a time rather than just one or two. iTunes manages your songs and videos. iPhoto and Picasa manage your photos, and Outlook and Apple Mail store your e-mails. The UI for aggregator applications is typically navigation based, consisting of top-level navigable categories in a left-side panel (playlists in iTunes, folders in Mail, albums in iPhoto), and scrolling listings in the main window.

❑ **Widgets:** A third category is "widget"-style applications, which are mini applications that display system or other information (battery status meter, weather, world clock) or perform a specific task (lyrics grabber, radio tuner). A widget UI typically consists of a single screen and a settings pane.

On the desktop, task-based applications have traditionally been the dominant category, although aggregators have become more and more important over the past decade with the increasing need to manage digital media. Although widgets are quite popular now that Apple and Microsoft have added this functionality directly into their OS, they remain far less important.

When you look at built-in iPhone applications, you can see that they generally fall into these three categories. However, because of iPhone's viewport and file storage constraints, task-based applications take a backseat role to the aggregators (see Table 4-1).

Table 4-1: Categorizing Apple's Built-In iPhone Applications

Aggregators	Task-based	Widgets
Mail	Safari	Stocks
Messages	Phone	Weather
Photos	Camera	Clock
YouTube	Calendar	Calculator
Notes	Maps	
Contacts (Address Book)	Compass	
iPod	Voice Memos	
iTunes		
App Store		

Although the document is the primary point of focus in a traditional desktop application, a document is often consumable and nonpermanent on the iPhone device. Most of the documents that users work with are consumable: Web pages, SMS messages, YouTube videos, quick notes, Google maps. Even Word, Excel, and Acrobat documents are read-only and are only accessible as e-mail attachments. What's more, for the more permanent storage pieces of information, you tend to sync with a master copy on your desktop — iPod songs, videos, and photos. In fact, there are only three cases in which you actually create data on the iPhone that you then store permanently — calendar appointments, e-mailed photos, and contacts.

The focus of iPhone usage is consuming information far more than creating information. If your application conforms to this usage model, your UI design needs to account for that reality.

Navigation List–Based UI Design

Because the focus of the iPhone is to consume various amounts of information, navigation list–based design becomes an essential way to present large amounts of information to users. As I mentioned earlier, desktop applications typically relegate navigation lists to a side panel on the left of the main window, but many iPhone applications use "edge-to-edge" navigation as the primary driver of the UI.

Not all navigation list designs are equal. In fact, the iPhone features at least eight distinct varieties of navigation lists. For example, the Contacts list uses a single line to display the name of a contact in bold letters (see Figure 4-5), whereas Mail uses a 4-line list style to display both message header information and optional text preview (see Figure 4-6). Finally, YouTube sports a wealth of information in its 4-line item (see Figure 4-7). Table 4-2 lists each of the various navigation style lists.

Figure 4-5: Contacts 1-line navigation list

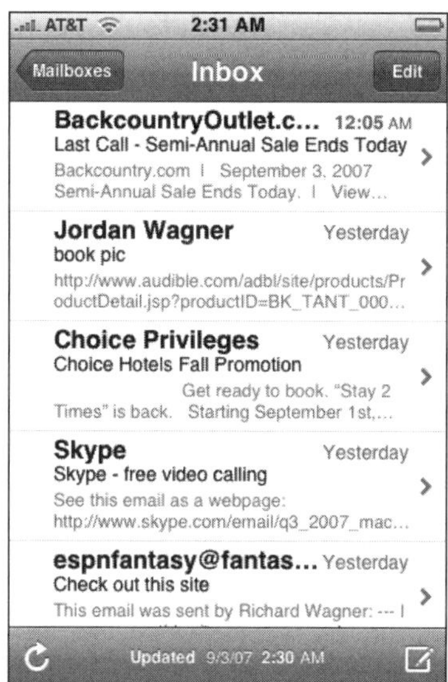

Figure 4-6: Mail's 4-line navigation list

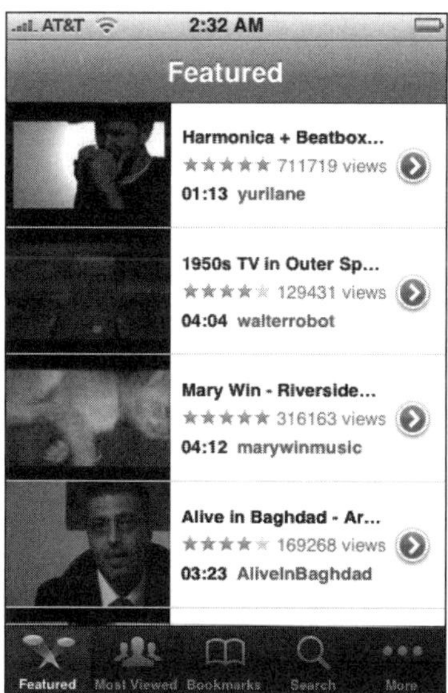

Figure 4-7: YouTube 4-line navigation list

Table 4-2: Different Types of Navigation Lists

Application	Style	Displays
Contacts	1 line	Name of contact (last name bolded)
Mail	2.7 lines (default 4)	Message title and optional text preview
Google Maps List	2 lines	Location name and address
SMS	3 lines	Message title and text preview
Photos	1 line	Album title and thumbnail image
YouTube	3 lines	Thumbnail, title, rating, length, views, and submitter
Notes	1 line	First line of note text
iPod Playlists	1 line	Playlist name
Settings	1 line	Grouped items with icons

However, no matter the style of the navigation lists, they are designed to quickly take you to a destination page in as few interactions as possible.

What's more, the top title bar of the app usually provides contextual information to help users understand where they are in the hierarchy of the application. The left side often has a Back button that enables users to return to the previous screen that they were on. More on this in the "Title Bar" section to come later.

Application Modes

Native iPhone applications also often have modes or views to the information or functionality with which you can work. These modes are displayed as icons or buttons on the bottom toolbar (see Figure 4-8). Interestingly, in the initial release of iCal, the buttons were on the top. However, in subsequent releases, Apple moved the buttons to the bottom to adhere to the convention of *navigation on top, modes on bottom* (see Figure 4-9).

Table 4-3 details these modes.

Table 4-3: Application Modes and UI Access

Application	Modes	UI Controls
iCal	List, Day, Month	Bottom button bar
Phone	Favorites, Recents, Contacts, Keypad, Voicemail	Bottom toolbar
iPod	Playlists, Podcasts, Albums, Videos, and so on	Bottom toolbar
YouTube	Featured, Most Viewed, Bookmarks, Search	Bottom toolbar
Clock	World Clock, Alarm, Stopwatch, Timer	Bottom toolbar

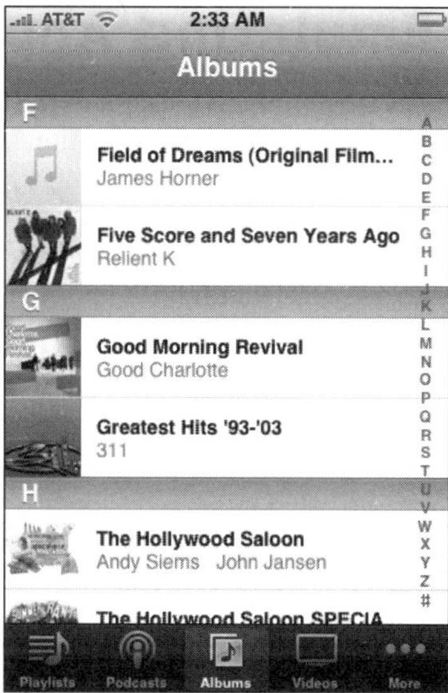

Figure 4-8: The bottom toolbar in iPod provides different views of a digital media library.

Figure 4-9: iCal now puts its calendar view toggle buttons on the bottom.

Therefore, as you begin to examine how the UI of your application should be designed, look to see what parallels exist with the built-in iPhone application design, and emulate its general look and feel.

Screen Layout: Emulating Apple Design

By the time you have studied and evaluated the UI design of the built-in applications, you can begin to determine what parallels may exist with the type of application in which you are building.

For applications that need to use a navigation list design, download one of the frameworks that I discussed back in Chapter 3, "Building with Web App Frameworks." Each of these enables you to easily implement edge-to-edge navigation list–based applications.

The four components of a typical iPhone application are a title bar, a navigation list, a destination page, and a button bar.

The Title Bar

Most iPhone Web applications will want to include a title bar to emulate the look of the standard title bar available in nearly all native iPhone applications. When the URL bar is hidden (and I explain how to do this later in this chapter), the custom title bar appears just below the status bar at the top of the viewport (see Figure 4-10).

The title bar includes the following elements:

❑ **Back button:** A Back button should be placed on the left side of the toolbar to allow the user to return to the previous page. The name of the button should be the same as the title of the previous screen. This "bread crumbs" technique lets users know how they got to the page and how to get back. If the page is at the top level of the application, remove the Back button completely.

❑ **Screen title:** Each screen should have a title displayed in the center of the toolbar. The title of the page should be one word and appropriately describe the content of the current screen. You will not want to include the application name in each screen title of the application, as you will for a standard Web application.

❑ **Command button:** For some screens, you will want to employ a common command, such as Cancel, Edit, Search, or Done. If you need this functionality, place a command button at the top right of the title bar.

Edge-to-Edge Navigation Lists

If your application aggregates or organizes lists of information, you will typically want your UI to emulate iPhone's edge-to-edge navigation list design, as shown in Figure 4-11. Each of the cells, or subsections, is extra large to allow for easy touch input. In addition, to ensure that users never lose context and get lost, the title shows the current page, while a Back button indicates the screen to which users can return if they choose. And, when a list item expands to a destination page or another list, an arrow is placed on the right side indicating a next page is available to the right.

When a list item is selected, the navigation list should emulate Apple's slide-in animation, appearing as if the new page is coming in from the right side of the screen, replacing the old.

Table 4-4 lists each of the specific metrics to emulate the same look and feel of the Apple design in edge-to-edge navigation lists. Note that iUI defines navigation lists based on these specifications and implements the slide-in animation effect.

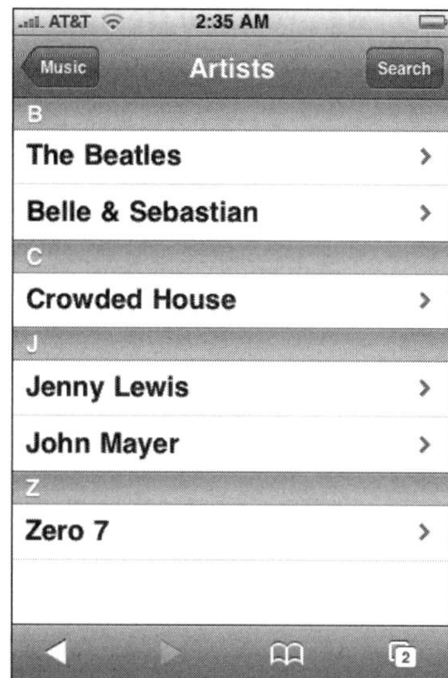

Figure 4-10: Title bar

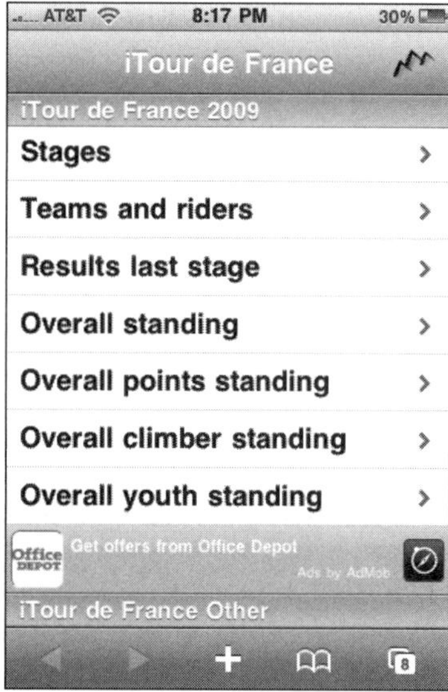

Figure 4-11: Emulating Apple's edge-to-edge navigation design

Table 4-4: Metrics for Apple's Edge-to-Edge Design

Item	Value
Cell height (including bottom line)	44px
Cell width	320px (portrait), 480px (landscape)
Font	Helvetica, 20pt bold (normal text acceptable for less important text)
Font color	Black
Horizontal lines (between cells)	#d9d9d9 (RGB=217, 217, 217)
Left padding	10px
Bottom padding	14px
Control height	29px
Control alignment	Right, 10px
Control shape	Rounded Rectangle of 7-degree radius
Control text	Helvetica, 12pt
Background color	White

Rounded Rectangle Design Destination Pages

In a navigation list UI design, users will ultimately wind up at a destination page that provides a full listing of the specific piece of information in which they were looking. Apple implements a rounded rectangle design, as shown in Figure 4-12. Labels are displayed on a blue background, while items are grouped logically and surrounded by a rounded rectangle box. Table 4-5 describes the specifications you should follow to implement this Apple design.

Table 4-5: Metrics for Apple's Rounded Rectangle Design

Item	Value
Cell height	44px
Rounded rectangle corner radius	10px × 10px radius (`-webkit-border-radius:10px`)
Rounded rectangle left and right margins	10px
Rounded rectangle top and bottom margins	17px
Horizontal lines (between cells)	#d9d9d9 (RGB=217, 217, 217)
Label font	Helvetica 17pt, bold

Item	Value
Label font color	#4c566c (RGB=76, 86, 108)
Cell font	Helvetica 17pt, bold
Cell font color	Black
Cell text position	10px from left edge, 14px bottom edge
Background color	#c5ccd3 (RGB= 197, 204, 211)

Figure 4-12: Implement rounded rectangle design for destination pages

The Button Bar

If you show your app in full-screen mode (which I talk about later), you can also take advantage of the added real estate by emulating the "mode bar" (also called a *tab bar*).

Designing for Touch

One of the most critical design considerations you need to take into account is that you are designing an interface that will interact with a finger, not a mouse or other mechanical pointing device. Whereas a mouse pointer has a small point just a couple pixels in height, a finger can touch 40 pixels or more of the screen during a typical click action. Therefore, when laying out controls in an application, make sure the height of controls and spacing between controls are easy to use even for someone with large fingers.

Because iPhone is a mobile device, keep in mind that users may be on the go when they are interacting with your application. Maybe they are walking down the street, waiting in line at Starbucks, or perhaps even jogging. Therefore, you will want to allow enough space in your UI to account for shaky fingers in those use case scenarios.

Standard navigation list cells should be 44px in height. Buttons should be sized about the size of a finger — typically 40px in height or more — and have sufficient space around them to prevent accidental clicks. You can get by with a button of 29–30 pixels in height if no other buttons are around it, but be careful. Table 4-6 lists the recommended sizes of the common elements.

In addition to sizing and spacing issues, another important design decision is to minimize the need for text entry. Use select lists rather than input fields where possible. What's more, use cookies to remember last values entered to prevent constant data reentry.

Table 4-6: Metrics for Touch Input Screen

Element Metric	Recommended Size
Element height	40px (min. 29px)
Element width	Min. 30px
Select, Input height	30px
Navigation list cell height	44px
Spacing between elements	20px

Working with Fonts

With its 160 pixels-per-inch display and anti-aliasing support, the iPhone is an ideal platform to work with typefaces. Quality fonts render beautifully on the iPhone display, enhancing the overall attractiveness of your application's UI.

Helvetica, Apple's font of choice for the iPhone, should generally be the default font of your application. However, the iPhone does offer several font choices for the developer. Unlike a typical Web environment in which you must work with font families, the iPhone allows you make some assumptions on the exact fonts that users will have when they run your application. Figure 4-13 shows the fonts that are supported on the iPhone.

Safari automatically substitutes three unsupported fonts with their built-in counterparts. Courier New is substituted when Courier is specified. Helvetica is substituted for Helvetica Neue, and Times New Roman is used in place of Times.

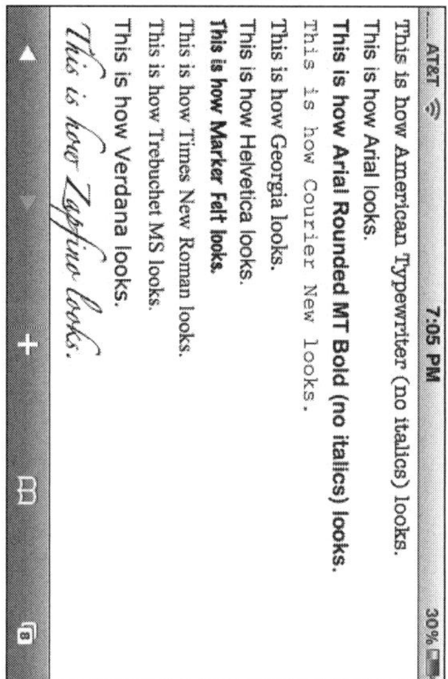

Figure 4-13: iPhone fonts

Best Practices in iPhone UI Design

When you are designing for the iPhone, you should keep in mind several best practices:

❑ **Remember the touch!** Perhaps no tip is more critical in iPhone UI design than double-checking every design decision you make with the reality of touch input. For example, ESPN's Podcenter, shown in Figure 4-14, uses a UI that roughly simulates the Apple navigation list design. However, notice the rows are thinner, making it harder to touch the correct podcast item, especially if the user is walking or performing another physical activity.

❑ **Make sure you design your application UI to work equally well in portrait and landscape modes.** Some native applications, such as Mail, optimize their UI for portrait mode and ignore changes that users make to orientation. Third-party iPhone Web app developers do not have that same level of control. Therefore, any UI design you create needs to work in both orientation modes.

❑ **Avoid UI designs that require horizontal scrolling.** If your interface design requires users to scroll from side to side within a single display screen, change it. Horizontal scrolling is confusing to users and leaves them feeling disoriented within your application.

❑ **Keep your design simple.** As attractive as the iPhone interface is, perhaps its most endearing quality is its ease of use and simplicity. Your UI design should follow suit. Avoid adding complexity where you do not need to — either in functionality or design (see Figure 4-15).

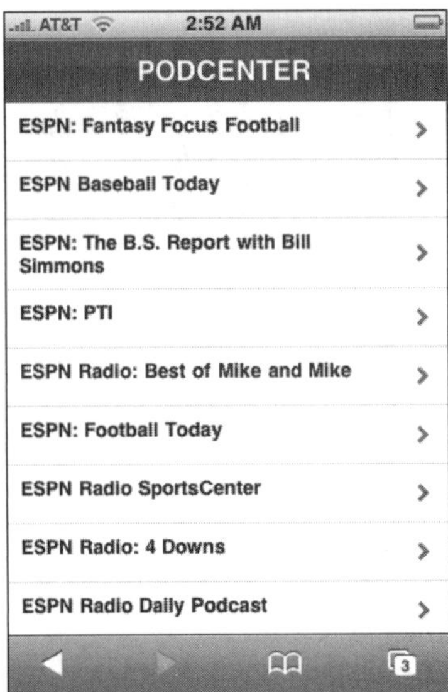

Figure 4-14: The shorter cells make it easy for fat or shaky fingers to select the wrong choice.

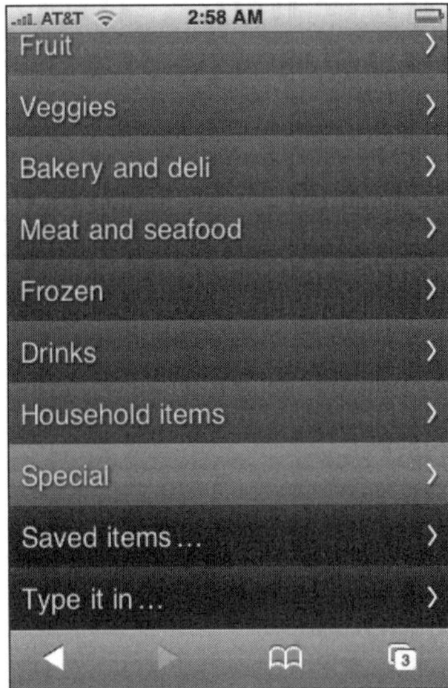

Figure 4-15: Multicolor list makes for hard reading

❑ **Use standard iPhone terminology.** You know the saying, "When in Rome." Well, when designing for the iPhone, be sure you do not bring along the UI baggage you are used to in the Windows, Mac, or the Web world. For example, "Preferences" are "Settings" and "OK" should be "Done."

❑ **Use UI frameworks, but use them wisely.** iPhone Web app frameworks (see Chapter 3) are major assets to the iPhone Web app developer community and provide a major head start in developing applications. However, don't automatically assume that their prebuild designs are the best way to go for your application. You may find another approach is better for your specific needs.

❑ **Restrict the use of a black button bar.** A translucent black button bar (such as the one used by the iPod app in Figure 4-8) should only be used in your application for displaying modes or views, not for commands. Use the other button types for commands.

❑ **Minimize the rabbit trail.** Because iPhone users are primarily concerned with consuming data, you will want to get them to their destination page as soon as possible. Therefore, make sure you are optimally organizing the information in a way that enables users to get to the data they need in just a couple of flicks and taps.

❑ **Place text entry fields at the top of the page.** When your application requires data entry fields, work to place these input fields as near to the top of the page as possible. Top positioning of text entry fields helps minimize the chances of the user losing context when the on-screen keyboard suddenly appears at the bottom of the screen when the user selects the field.

❑ **Communicate status.** Because your application may be running via a 3G or even an EDGE connection, its response may be relatively slow. As a result, be sure to provide status to the user when performing a function that requires server processing. This visual clue helps users feel confident that your application is working as expected and is not in a hung state.

❑ **Label the title bar appropriately.** Make sure each screen/page has its own title. The Back button should always be titled the same as the previous screen.

❑ **Unselect previously selected items.** When a user clicks the Back button in a navigation-list UI, be sure that the previously selected item is unchecked.

❑ **Break the rules — competently.** Although you should generally adhere to the Apple UI design guidelines that I've been discussing in this chapter, not every iPhone application UI needs to rigidly conform to a design implemented already by Apple. You may have an application in which a different look-and-feel works best for its target users. However, if you decide to employ a unique design, be sure it complements overall iPhone design, not clashes with it.

Finishing Touches: Making It Look Like a Native App

To transform your Web app into something that looks and feels like a native iPhone app, you need to put some finishing touches onto it. In this final section, I will show you how to launch the app in full-screen, customize the status bar, and create a WebClip icon.

Launching in Full-Screen Mode

By default, when you run your Web app in Safari on iPhone, Safari still takes up a sizable majority of the viewport with its URL bar and bottom bar. However, to take back that wasted space and to make your app look more like a native app, you'll want to launch your application in full-screen mode (see Figure 4-16).

Figure 4-16: Launching an app in
full-screen mode

To do so, add an `apple-mobile-web-app-capable` meta tag to the document head and assign it a `yes` value. For example:

```
<meta name="apple-mobile-web-app-capable" content="yes" />
```

Note that full-screen mode does not take effect unless you launch your app from the Home screen via a WebClip icon (see the later section titled "Adding a WebClip Icon"). When you access the page from the URL bar, Safari retains its URL bar and button bar.

Customizing the Status Bar

When you use the `apple-mobile-web-app-capable` meta tag to go into full-screen mode, Safari still displays the top status bar. This behavior is not that unexpected, because most native apps show it in the same way. However, if you would prefer to mark the status bar, you can customize its appearance to be transparent black by using the `status-bar-style` meta tag and setting it to a value of `black`. For example:

```
<meta name="apple-mobile-web-app-status-bar-style" content="black" />
```

Note that this meta tag only takes effect if you also use the `apple-mobile-web-app-capable` meta tag to turn on full-screen mode.

Figure 4-17 shows a customized status bar.

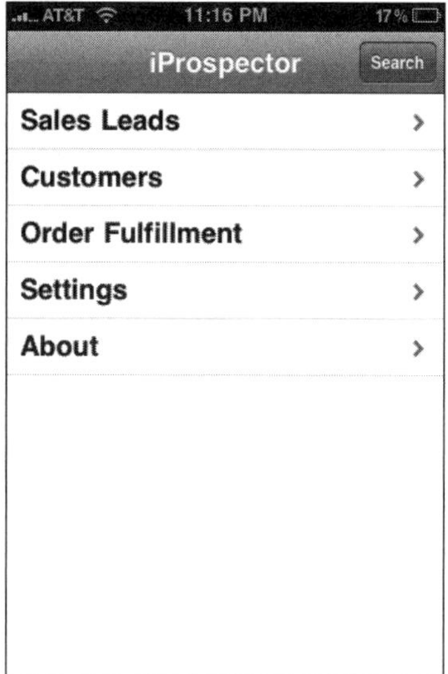

Figure 4-17: Customizing the status bar

Adding a WebClip Icon

Web apps can be added to the home screen just like a native application. Developers can specify custom WebClip icons by adding the following link tag to the head of an HTML document:

```
<link rel="apple-touch-icon" href="apple-touch-icon.png"/>
```

The graphic file you point to needs to adhere to the following conventions:

❑ Be in PNG format.

❑ Be located in the root directory.

❑ Measure 57 × 57 pixels (different sizes are scaled to 57 ? 57).

❑ Be a rectangle — the OS automatically adds the rounded corners.

❑ Be named `apple-touch-icon.png` or `apple-touch-icon-precomposed.png`. If you use `apple-touch-icon.png`, the iPhone OS adds round corners, a drop shadow, and a "shiny coating" to the graphic. If you use `apple-touch-icon-precomposed.png`, the OS won't add these effects.

Figure 4-18 shows WebClip icons from three iPhone Web apps.

It's probably not recommended under most Web app circumstances, but if you want to assign a WebClip icon for a specific page in your Web app, you can use a custom PNG name instead. For example:

```
<link rel="apple-touch-icon" href="page_icon.png"/>
```

Figure 4-18: WebClip icons on the
home screen

Users can add the WebClip icon of your app by clicking the + button on the button bar when they navigate to the URL. In the popup dialog, users click the Add to Home Screen button (see Figure 4-19).

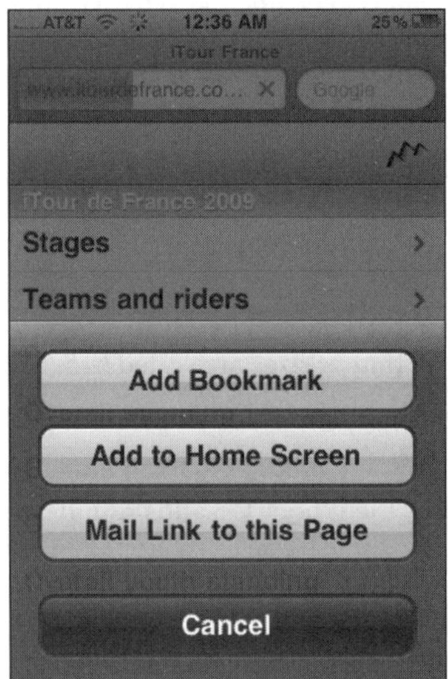

Figure 4-19: Add to Home Screen button

Unfortunately, because there is no formal install process of a Web app, there is no programmatic way to automatically add a WebClip icon to the home screen. The user must carry out the process.

Summary

This chapter focused on creating a navigable user interface for iPhone web apps. It began with a discussion of the viewport and how designing for it is different than a traditional web site. The chapter then explored core concepts and styles of native iPhone app UIs, providing exact measurements, sizes, and colors to utilize when you plan to create a user interface to emulate standard iPhone UI conventions. It continued by providing a list of best practices to consider when you design your web app. By following these tips and strategies, you'll avoid common pitfalls that web app developers can fall into when they initially develop for the iPhone platform. Finally, the chapter closed with a look at how to add important finishing touches that give an added level of professionalism.

5

Styling with CSS

Like its Mac and Windows cousins, Safari for iPhone provides some of the best CSS support of all Web browsers. As you develop iPhone Web applications, you can utilize CSS to make powerful user interfaces.

Safari offers support for CSS 2.1 as well as CSS 3. However, Safari also supports some properties technically called "experimental CSS 3" that are not currently part of the World Wide Web Consortium (W3C) CSS standard but will be supported by Apple going forward. (A `-webkit-` prefix is added to the names of these properties.) For a normal Web application, developers typically stay away from these experimental properties or at least do not rely on them for their application's design. However, because you know that an iPhone and iPod touch user will be using Safari/ WebKit, you can safely use these more advanced styles as you create your UI.

CSS Selectors Supported in Safari

Many would contend that the real power of CSS is not so much in the properties that you can apply but in CSS's ability to select the exact elements within a Document Object Model (DOM) that you want to work with. If you have worked with CSS before, you are probably quite familiar with the standard type, class, and ID selectors. However, Safari's support for selectors includes many new selectors that are part of the CSS3 specification. Table 5-1 lists a set of CSS selectors that Safari supports, and Table 5-2 lists the set of pseudoclasses and pseudoelements that Safari works with.

Note that the following CSS3 selectors are not supported with Safari:

- `:last-child`
- `:only-child`
- `nth-child()`
- `nth-last-child()`
- `last-of-type`

- ❏ `only-of-type`
- ❏ `:nth-of-type()`
- ❏ `:nth-last-of-type()`
- ❏ `Empty`

Table 5-1: Safari CSS Selectors

Selector	Definition
`E`	Type selector
`.class`	Class selector
`#id`	ID selector
`*`	Universal selector (all elements)
`E F`	Descendant selector
`E > F`	Child selector
`E + F`	Adjacent sibling selector
`E ~ F`	Indirect adjacent selector[a]
`E[attr]`	`attr` is defined
`E[attr=val]`	`attr` value matches `val`
`E[attr~=val]`	One of many attribute value selectors[b]
`E[attr|=val]`	`attr` value is a hyphen-separated list and begins with `val`[b]
`E[attr^=val]`	`attr` value begins with `val`[a,b]
`E[attr$=val]`	`attr` value ends with `val`[a,b]
`E[attr*=val]`	`attr` value contains at least one instance of `val`[a,b]

[a]New to CSS3

[b]Case sensitive, even when unnecessary

Table 5-2: Safari Pseudoclasses and Pseudoelements

Pseudoclass/ Pseudoelement	Definition
`E:link`	Unvisited link
`E:visited`	Visited link
`E:lang([Code])`	Selector content uses the language code specified
`E:before`	Content before an element

Pseudoclass/ Pseudoelement	Definition
E::before	Content before an element (new double-colon notation in CSS3)[a]
E:after	Content after an element
E::after	Content after an element (new double-colon notation in CSS3)[a]
E:first-letter	First letter of element
E::first-letter	First letter of element (new double-colon notation in CSS3)[a]
E:first-line	First line of element
E::first-line	First line of element (new double-colon notation in CSS3)[a]
E:first-child	First child[b]
E:first-of-type	First child of type[a,b]
E:root	Root[a]
E:not()	Negation[a]
E:target	Target[a]
E:enabled	Enabled state[a]
E:disabled	Disabled state[a]
E:checked	Checked state[a]

[a]New to CSS3

[b]When new first child/child of type is created programmatically using JavaScript, the previous child maintains the :first-child or :first-of-type attributes.

Text Styles

When you are styling text inside your iPhone Web applications, keep in mind three text-related styles that are important to effective UI design: -webkit-text-size-adjust, text-overflow, and text-shadow. These properties are explained in this section.

Controlling Text Sizing with -webkit-text-size-adjust

When a page is rendered, Safari automatically sizes the page's text based on the width of the text block. However, by using the -webkit-text-size-adjust property, you can override this setting. The none option turns off auto-sizing of text:

```
body { -webkit-text-size-adjust: none; }
```

Or, you can specify a specific multiplier:

```
body { -webkit-text-size-adjust: 140%; }
```

Finally, you can set it to the default value of `auto`:

```
body { -webkit-text-size-adjust: auto; }
```

Figures 5-1, 5-2, and 5-3 show the results of these three options on the same page.

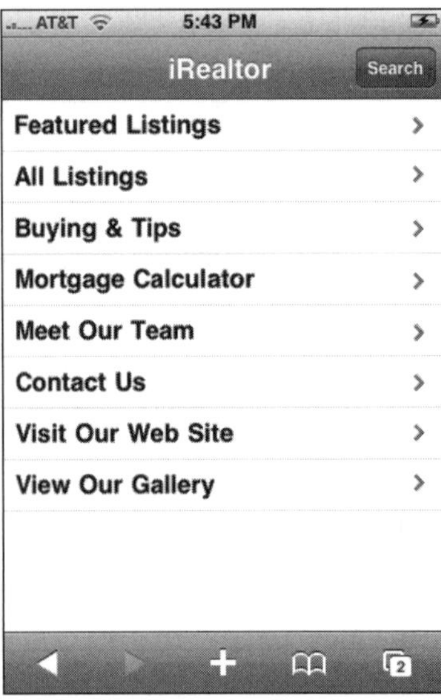

Figure 5-1: No text adjustment

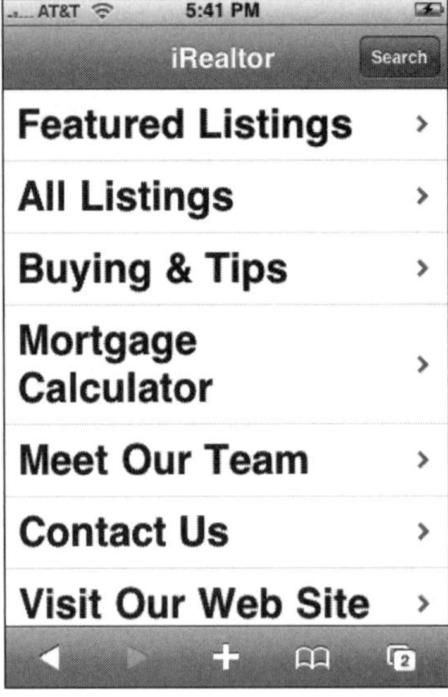

Figure 5-2: Text is increased to 140%

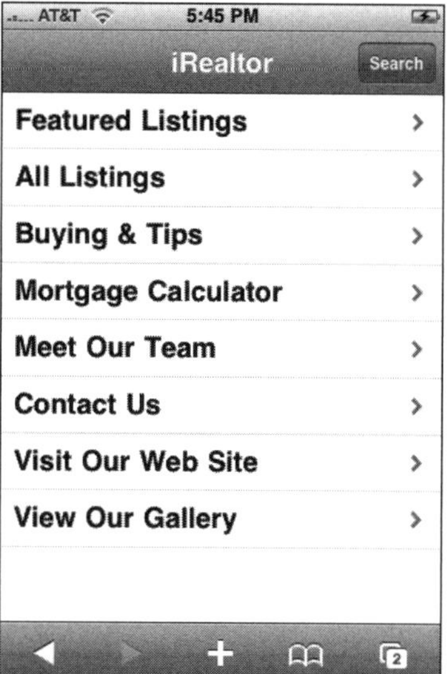

Figure 5-3: Text is adjusted based on the width of the content block.

For a normal Web site, `-webkit-text-size-adjust: auto` is recommended for improving the readability of text. However, if you are developing an application, you will almost always want to use `-webkit-text-size-adjust: none` to maintain precise control over the text sizing, particularly when you go between portrait and landscape modes.

Handling Overflowed Text with text-overflow

Because the width of the viewport in Safari is either 320 (portrait) or 480 (landscape) pixels, effectively managing the physical length of dynamic text on UI elements can be tricky. This is particularly important for headings or button text in which a fixed amount of real estate is available. The best example of the need to handle text overflow is in the top toolbar that is a standard part of the iPhone application interface. By default, any content that does not fit inside the container box of the element is clipped, which can potentially lead to confusion, such as the back button example shown in Figure 5-4. Because there is not enough space to display the text `iProspector`, only `iProspect` is shown.

To prevent this situation from happening, you can provide a visual hint that the text has been clipped. Fortunately, the `text-overflow` property enables developers to specify what they want to have done when the text runs on. The two values are `ellipsis` and `clip`. The `ellipsis` value trims the content and adds an ellipsis character (...) to the end. Suppose you assign the following property to the toolbar's button and heading element:

```
text-overflow: ellipsis;
```

Now, when text overflows, an ellipsis is added, as shown in Figure 5-5.

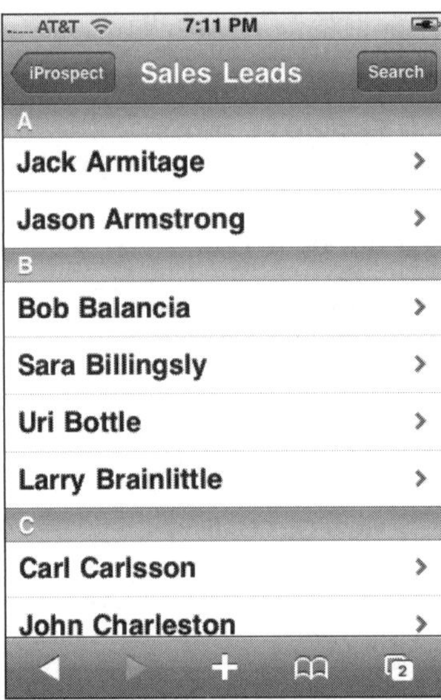

Figure 5-4: Text is clipped if it does not fit into the available space.

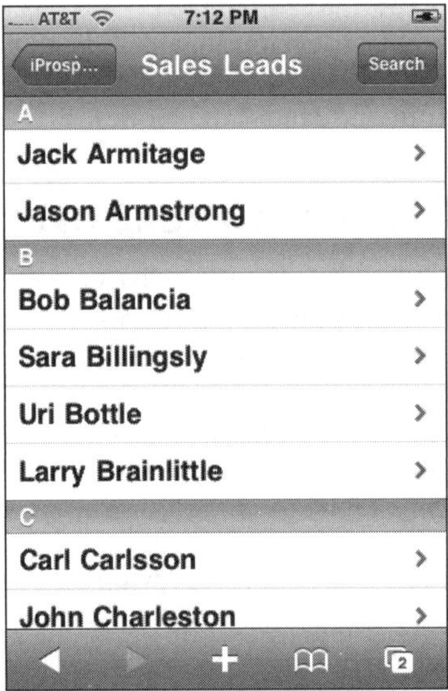

Figure 5-5: An ellipsis provides a visual indicator that the text has been clipped.

The `text-overflow` property is particularly useful for iPhone Web apps because a heading that displays fully in landscape mode may need to be clipped in the much thinner portrait mode.

The use of `text-overflow` may require specifying additional CSS properties to display as intended. The following code, for example, needs to have `overflow` and `white-space` properties set to ensure that the `text-overflow` property works:

```
<html>
<meta name="viewport" content="width=320; initial-scale=1.0; maximum-scale=1.0;">
<style>
.ellipsis
{
    text-overflow: ellipsis;
    width: 200px;
    white-space: nowrap;
    overflow: hidden;
}

.ellipsisBroken1
{
    text-overflow: ellipsis;
    width: 200px;
    /* white-space: nowrap; */
    overflow: hidden;
}

.ellipsisBroken2
{
    text-overflow: ellipsis;
    width: 200px;
    white-space: nowrap;
    /* overflow: hidden; */
}
</style>
<body>
<div class="ellipsis"> this is a test this is a test this is a test
this is a test this is a test this is a test this is a test </div>
<br><br>
<div class="ellipsisBroken1"> this is a test this is a test this is a test
this is a test this is a test this is a test this is a test </div>
<br><br>
<div class="ellipsisBroken2"> this is a test this is a test this is a test
this is a test this is a test this is a test this is a test </div>
</body>
</html>
```

Subtle Shadows with text-shadow

In the iPhone UI, Apple makes subtle use of text shadows, particularly on buttons and larger heading text. Besides lending to aesthetics, text shadows are useful in making text more readable by increasing its contrast with the background.

You can add drop shadows to your text through the `text-shadow` property. The basic declaration is as follows:

```
text-shadow: color offsetX offsetY blurRadius;
```

The first value is the color of the shadow. The next two give the shadow's offset position — the second value being the *x*-coordinate and the third being the *y*-coordinate. (Negative values move the shadow left and up.) The fourth parameter indicates the shadow's Gaussian blur radius. So, in the following example, a gray shadow is added 1px above the element's text with no blur:

```
text-shadow: #666666 0px -1px 0;
```

However, text shadows can be distracting and look tacky if they are too noticeable. Therefore, an `rgba` (red, green, blue, alpha) color value can be used in place of a solid color value to define the transparency value of the shadow. (See the "Setting Transparencies" section later in this chapter.) Therefore, the following declaration defines a white shadow with a `0.7` alpha value (`0.0` is fully transparent, whereas 1.0 is fully opaque) that is positioned 1 pixel under the element's text:

```
text-shadow: rgba(255, 255, 255, 0.7) 0 1px 0;
```

Styling Block Elements

You can quickly apply several styles to block elements to transform their appearance and go beyond ordinary CSS styles that you may have used on traditional Web sites. These include three so-called experimental properties (`-webkit-border-image`, `-webkit-border-radius`, and `-webkit-appearance`) and a CSS3 enhancement of the `background` property. These are described in this section.

Image-Based Borders with -webkit-border-image

The `-webkit-border-image` property enables you to use an image to specify the border rather than the `border-style` properties. The image appears behind the content of the element, but on top of the background. For example:

```
-webkit-border-image: url(image.png) 7 7 7 7;
```

The four numbers that follow the image URL represent the number of pixels in the image that should be used as the border. The first number indicates the height of the top (both the corners and edge) of the image used. Per CSS conventions, the remaining three numbers indicate the right, bottom, and left sides. Pixel is the default unit, although you can specify percentages.

If the image URL you provide cannot be located or the style is set to `none`, `border-style` properties are used instead.

Optionally, you can specify one or two keywords at the end of the declaration. These keywords determine how the images for the sides and the middle are scaled and tiled. The valid keywords are stretch or round. If stretch is used as the first keyword, the top, middle, and bottom parts of the image are scaled to the same width as the element's padding box. Far less common for iPhone use, round can also be used as the first keyword. When this setting is present, the top, middle, and bottom images are reduced in width so that a whole number of the images fits in the width of the padding box. The second keyword acts on the height of the left, middle, and right images. If both keywords are omitted, stretch stretch is implied.

When rendered, Safari looks at the -webkit-border-image property and divides the image based on the four numbers specified.

The -webkit-border-image property plays an important role in creating CSS-based iPhone buttons, which is explained later in this chapter.

Rounded Corners with -webkit-border-radius

The -webkit-border-radius specifies the radius of the corners of an element. Using this property, you can easily create rounded corners on your elements rather than resorting to image-based corners. For example:

```
-webkit-border-radius: 10px;
```

This declaration specifies a 10px radius for the element, which is the standard radius value for the Rounded Rectangle design for destination pages. You can also specify the radius of each corner using the following properties:

```
-webkit-border-top-left-radius
-webkit-border-top-right-radius
-webkit-border-bottom-left-radius
-webkit-border-bottom-right-radius
```

If, for example, you wanted to create a div with rounded top corners but square bottom corners, the style code would look like the following:

```
div.roundedTopBox
{
    -webkit-border-top-left-radius: 10px;
    -webkit-border-top-right-radius: 10px;
    -webkit-border-bottom-left-radius: 0px;
    -webkit-border-bottom-right-radius: 0px;
}
```

Results are shown in the text box shown in Figure 5-6.

Figure 5-6: Rounded top, square bottom.

Gradient Push Buttons with -webkit-appearance

The -webkit-appearance property is designed to transform the appearance of an element into a variety of different controls. Safari for iPhone supports just two of the possible values: push-button and button. The push-button property holds the most promise for iPhone application developers. Suppose, for example, you would like to turn a link element into a gradient push button. You could do it with an image, but -webkit-appearance: push-button allows you to do it entirely within CSS. To demonstrate, begin with a link assigned to a class named special:

```
<a href="tel:202-555-1212" class="special">Call Headquarters</a>
```

Then define the a.special style:

```
a.special
{
    display: block;
    width: 246px;
    font-family: Helvetica;
    font-size: 20px;
    font-weight: bold;
    color: #000000;
    text-decoration: none;
    text-shadow: rgba(255, 255, 255, 0.7) 0 1px 0;
    text-align: center;
    line-height: 36px;
    margin: 15px auto;
    -webkit-border-radius:10px;
    -webkit-appearance: push-button;
}
```

The `display:block` and `width:246px` properties give the link a wide rectangular block shape. The `-webkit-appearance: push-button` property transforms the appearance to a gradient gray push button. The `-webkit-border-radius` rounds the edges using the standard 10px value. Although the shape of the push button is now set, the text needs to be tweaked using not just standard text formatting properties, but also a `line-height` property of 36px, which vertically centers the 20px text in the middle of the push button. If you add a simple `background-color: #999999` style to the `body` tag, you get the result shown in Figure 5-7.

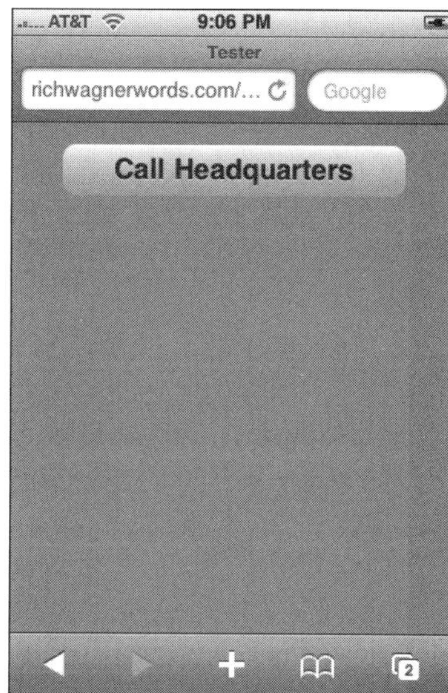

Figure 5-7: Gradient push button

Multiple Background Images

In earlier versions of CSS, there was always a 1:1 correspondence between an element and a background image. Although that capability worked for most purposes, some page designs could not work effectively with a single background image defined. So, to get around the 1:1 limitation, designers would resort to adding extra `div` tags here or there just to achieve the intended visual design.

CSS3 addresses this issue by allowing you to define multiple background images for a given element. Most browsers don't support this feature yet, but fortunately for iPhone Web application developers, Safari/WebKit does.

You define a set of background images by listing them in order after the `background` property name declaration. Images are rendered with the first one declared on top, the second image behind the first, and so on. You can also specify the `background-repeat` and `background-position` values for each of the images. If `background-color` is defined, this color is painted below all the images. For example:

```
div.banner
{
    background: url(header_top.png) top left no-repeat,
      url(banner_main.png) top 6px no-repeat,
```

```
        url(header_bottom.png) bottom left no-repeat,
        url(middle.png) left repeat-y;
    }
```

In this code, `header_top.png` serves as the background image aligned to the top-left portion of the `div` element. `banner_main.png` is positioned 6px from the top, whereas `header_bottom.png` is positioned at the bottom of the `div`. Finally, `middle.png` is treated as a repeating background.

Setting Transparencies

Developers have long used `rgb` to specify an RGB color value for text and backgrounds. CSS3 adds the ability to set an alpha value when specifying an RGB color with the new `rgba` declaration. Using the `rgba` declaration, you can add translucent color overlays without transparent PNGs or GIFs. The syntax follows:

```
rgba(r, g, b, alpha)
```

The `r`, `g`, and `b` values are integers between 0 and 255 that represent the red, green, and blue values. `alpha` is a value between 0 and 1. (0.0 is fully transparent, whereas 1.0 is fully opaque). For example, to set a red background with a 50 percent transparency, you would use the following:

```
background: rgba(255, 0, 0, 0.5);
```

Keep in mind that the alpha value in the `rgba` declaration is not the same as the `opacity` property. `rgba` sets the opacity value only for the current element, whereas `opacity` sets the value for the element and its descendants.

The following example shows five `div` elements, each with a different `alpha` value for the black background:

```
<!DOCTYPE html PUBLIC "-//W3C//DTD XHTML 1.0 Strict//EN"
        "http://www.w3.org/TR/xhtml1/DTD/xhtml1-strict.dtd">
<html xmlns="http://www.w3.org/1999/xhtml">
<head>
<title>RGBA Declaration</title>
<meta name="viewport" content="width=320; initial-scale=1.0;
maximum-scale=1.0; user-scalable=0;">
<style type="text/css" media="screen">
div.colorBlock {
    width: 50px;
    height: 50px;
    float: left;
    margin-bottom: 10px;
    font-family: Helvetica;
    font-size: 20px;
    text-align:center;
    color:white;
    text-shadow: rgba(0,0, 0, 0.7) 0 1px 0;
    line-height: 46px;
}
```

```
</style>
</head>
<body>
<div style="margin: 10px 0 0 30px;">
<div class="colorBlock" style="background: rgba(0, 0, 0, 0.2);"><span
>20%</span></div>
<div class="colorBlock" style="background: rgba(0, 0, 0,
0.4);"><span>40%</span></div>
<div class="colorBlock" style="background: rgba(0, 0, 0,
0.6);"><span>60%</span></div>
<div class="colorBlock" style="background: rgba(0, 0, 0,
0.8);"><span>80%</span></div>
<div class="colorBlock" style="background: rgba(0, 0, 0,  1.0)
;"><span>100%</span></div>
</div>
</body>
</html>
```

Figure 5-8 shows the page in Safari.

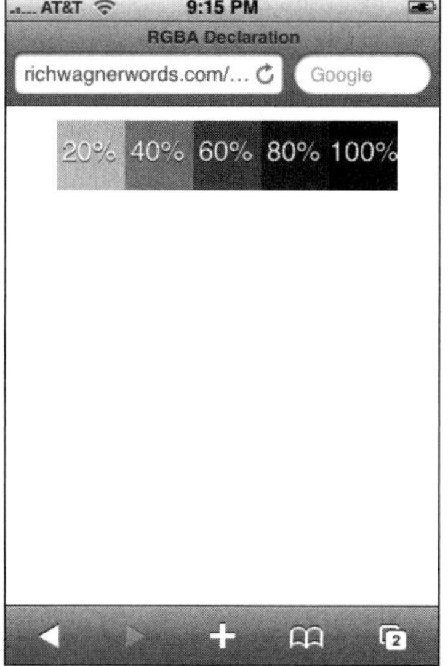

Figure 5-8: Alpha values in the rgba
declaration

Creating CSS-Based iPhone Buttons

Using `-webkit-border-image`, you can create push buttons that closely emulate Apple's standard push button design. This technique, inspired by developer Matthew Krivanek, involves using a pill-shaped button image (available for download at `www.wrox.com`), stretching the middle of the button image, but ensuring that the left and right sides of the button are not distorted in the process.

Begin by defining a normal link with a `fullSizedButton` class:

```
<a href="mailto:rich@digitalwalk.net" class="fullSizedButton">Send to Client</a>
```

Next, define the `a.fullSizedButton` style:

```
a.fullSizedButton
{
        font-family: Helvetica;
        font-size: 20px;
        display: block;
        width: 246px;
        line-height: 46px;
        margin: 15px auto;
        text-align:center;
        text-decoration: none;
        font-weight: bold;
        color: #000000;
        text-shadow: rgba(255, 255, 255, 0.7) 0 1px 0;
        border-width: 0 14px 0 14px;
        -webkit-border-image: url(images/whiteButton.png) 0 14 0 14;
}
```

In the preceding code, the `display` property is set to `block` and the width is set to 246px, the width of the buttons Apple uses. The `line-height` is set to 46px, which gives the block element the standard height and vertically centers the button text. A `border-width` property sets the left and right borders to 14px and eliminates the borders for the top and bottom by defining their values as `0`.

Now that everything else is set up, look at the `-webkit-border-image` property definition. In this example, 0 pixels are used from `whiteButton.png` on the top and bottom. However, the first 14 pixels of the image are used for the left border of the element, whereas the 14 rightmost pixels are used for the right border. Because the `whiteButton.png` image is 29 pixels in width, a 1-pixel section is used as the middle section. This middle section is then repeated over and over to fill the width of the element. Figure 5-9 shows how `-webkit-border-image` divides the image.

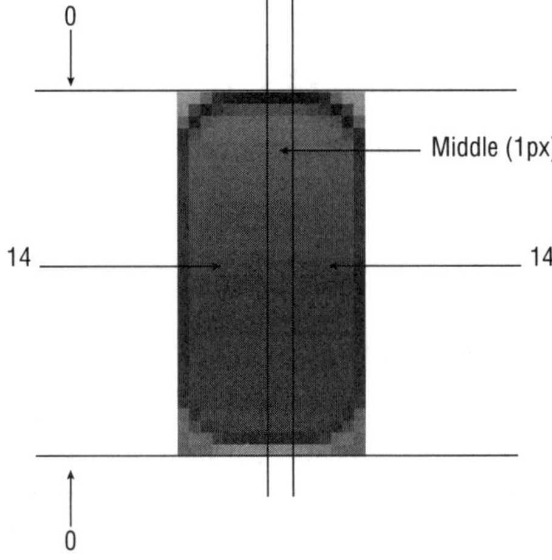

Figure 5-9: Carving up an image for a border

Figure 5-10 shows the button when rendered by Safari.

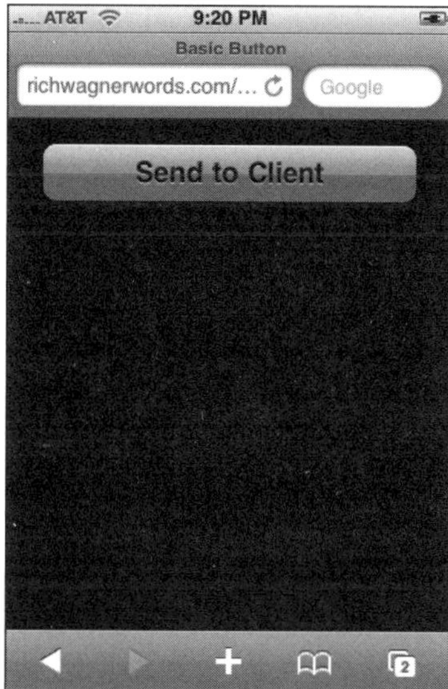

Figure 5-10: Using border-image to
style a button

Here is the full source code for this example:

```
<!DOCTYPE html PUBLIC "-//W3C//DTD XHTML 1.0 Strict//EN"
        "http://www.w3.org/TR/xhtml1/DTD/xhtml1-strict.dtd">
<html xmlns="http://www.w3.org/1999/xhtml">
<head>
<title>Basic Button/title>
<meta name="viewport" content="width=320; initial-scale=1.0;
maximum-scale=1.0; user-scalable=0;">
<style type="text/css" media="screen">
a.fullSizedButton
{
    font-family: Helvetica;
    font-size: 20px;
    display: block;
    width: 246px;
    margin: 15px auto;
    text-align:center;
    text-decoration: none;
    line-height: 46px;
    font-weight: bold;
    color: #000000;
    text-shadow: rgba(255, 255, 255, 0.7) 0 1px 0;
    border-width: 0 14px 0 14px;
    -webkit-border-image: url(images/whiteButton.png) 0 14 0 14;
}
body
```

```
{
     background-color: black;
}
</style>
</head>
<body>
<a href="mailto:me@company.net" class="fullSizedButton">Send to Client</a>
</body>
</html>
```

Identifying Incompatibilities

Although Safari for iPhone is closely related to its Mac and Windows counterparts, it is not identical in terms of CSS support. The latest versions of Safari for Mac and Windows support most of the newer CSS3 and experimental properties (prefixed with `-webkit-`). Safari for iPhone, however, provides limited support of several properties.

The following CSS properties are not supported (or have limited support) in Mobile Safari:

- `box-shadow`
- `-webkit-box-shadow`
- `text-stroke`
- `-webkit-text-stroke`
- `text-fill-color`
- `-webkit-text-fill-color`
- `-website-appearance` (`push-button` supported, but no other values are)

Summary

In this chapter, I walked you through the details of styling your iPhone Web apps using CSS. I began by exploring the selectors that are supported by Safari on iPhone. Next, the chapter explored text styles, focusing on the issues that you'll encounter designing for the iPhone viewport, particularly text sizing and text shadows. I continued on with a discussion on styling block elements, showing how to use image-based borders, rounded corners, and transparencies to create an attractive user interface. Finally, I showed you how to create 3-D style buttons using CSS alone that emulate iPhone conventions.

6

Programming the Interface

The previous two chapters surveyed the UI standards and guidelines that you need to keep in mind as you design a Web application that works well on iPhone as well as iPod touch. With these design principles in hand, you are ready to apply them as you develop and program your Web app.

To demonstrate how to implement an iPhone interface, I will walk you through a case study application I am calling iRealtor. The concept of iRealtor is to provide a mobile *house-hunter* application for potential buyers. The current pattern for Internet-based house hunting is to search MLS listings online, print individual listing addresses, get directions, and then travel to these houses. However, with iRealtor, you can do all those tasks on the road with an iPhone-based application. The design goals of iRealtor are to provide a way for users to do the following:

- Browse and search the MLS listings of a local realtor.

- Get a map of an individual listing directly from its listing page.

- Access information about the realtor and easily contact the realtor using iPhone phone or mail services.

- Browse other helpful tools and tips.

As you look at these overall objectives, an edge-to-edge navigation design looks like an obvious choice given the task-based nature of the application. The realtor information will be relatively static, but the MLS listings need to be database-driven. Therefore, you will take advantage of Ajax to seamlessly integrate listing data into the application.

Here's an overview of the technologies used for iRealtor:

- XHTML/HTML and CSS for presentation layer

- JavaScript for client-side logic

- Ajax for loading data into the application

- PHP or other server-side technology to serve MLS listing data (not included in case study example)

As I walk you through the application, I'll examine both the custom code I am writing for iRealtor and the underlying styles and code that power it. Therefore, no matter the framework you decide to choose, you at least will have a solid grasp on the key design issues you need to consider.

Top Level of Application

The top level of iRealtor is best presented as an edge-to-edge navigation-style list that contains links to the different parts of the application. When assembled, the design will look like what is shown in Figure 6-1.

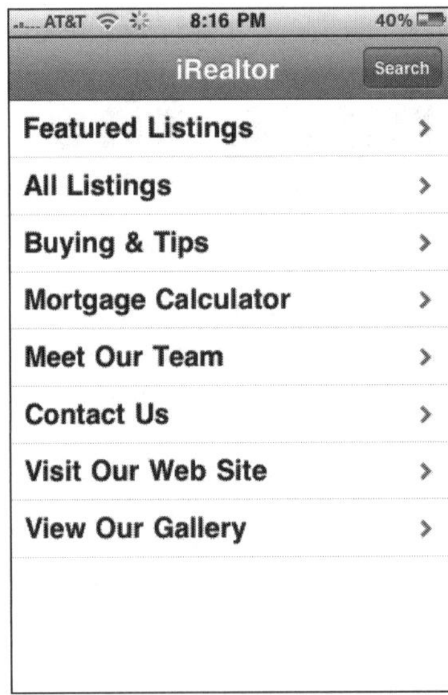

Figure 6-1: iRealtor top-level page

Creating index.html

To build the initial page, start with a basic XHTML document, linking the style sheet and scripting library files being used for this Web app:

```
<!DOCTYPE html PUBLIC "-//W3C//DTD XHTML 1.0 Strict//EN"
        "http://www.w3.org/TR/xhtml1/DTD/xhtml1-strict.dtd">
<html xmlns="http://www.w3.org/1999/xhtml">
<head>
<title>iRealtor</title> <meta name="apple-mobile-web-app-capable" content="yes" />
<meta name="viewport" content="width=device-width; initial-scale=1.0;
maximum-scale=1.0; user-scalable=0;" />
<style type="text/css" media="screen">@import "./iui/iui.css";</style>
<script type="application/x-javascript" src="./iui/iui.js"></script>
</head>
```

```
<body>
</body>
</html>
```

The `apple-mobile-web-app-capable` instruction opens the app in full screen when a user launches it from the Home screen. The viewport meta tag tells Safari exactly how to scale the page. It sets a 1.0 scale and does not change the layout on reorientation. It also specifies that the width of the viewport is the size of the device (`device-width` is a constant).

These properties ensure that iRealtor behaves like a Web application, not a Web page.

Examining Top-Level Styles

The style sheet sets up several top-level styles. The `body` style sets up the default `margin`, `font-family`, and `color`. It also uses `-webkit-user-select` and `-webkit-text-size-adjust` to ensure that iRealtor behaves as an application rather than a Web page. Here's the definition:

```
body {
    margin: 0;
    font-family: Helvetica;
    background: #FFFFFF;
    color: #000000;
    overflow-x: hidden;
    -webkit-user-select: none;
    -webkit-text-size-adjust: none;
}
```

For iPhone/iPod touch applications, it is important to assign `-webkit-text-size-adjust: none` to override the default behavior.

All elements, except for the `.toolbar` class, are assigned the following properties:

```
body > *:not(.toolbar) {
    display: none;
    position: absolute;
    margin: 0;
    padding: 0;
    left: 0;
    top: 45px;
    width: 100%;
    min-height: 372px;
}
```

In landscape mode, the `min-height` changes for these elements:

```
body[orient="landscape"] > *:not(.toolbar) {
    min-height: 268px;
}
```

The `orient` attribute changes when the orientation of the viewport changes between portrait and landscape. You'll see how this works later in the chapter.

The iUI framework uses a `selected` attribute to denote the current page of the application. From a code standpoint, the page is typically either a `div` or a `ul` list:

```
body > *[selected="true"] {
    display: block;
}
```

Links also are assigned the `selected` attribute:

```
a[selected], a:active {
    background-color: #194fdb !important;
    background-image: url(listArrowSel.png), url(selection.png) !important;
    background-repeat: no-repeat, repeat-x;
    background-position: right center, left top;
    color: #FFFFFF !important;
}
a[selected="progress"] {
    background-image: url(loading.gif), url(selection.png) !important;
}
```

The `a[selected="progress"]` style displays an animated GIF showing the standard iPhone loading animation.

Adding the Top Toolbar

The first UI element to add is the top toolbar, which serves a common UI element throughout the application. To create the toolbar, use a `div` element, assigning it the `toolbar` class:

```
<!--Top iUI toolbar-->
<div class="toolbar">
    <h1 id="pageTitle"></h1>
    <a id="backButton" class="button" href="#"></a>
    <a class="button" href="#searchForm">Search</a>
</div>
```

The `h1` element serves as a placeholder for displaying the active page's title. a `backbutton` is not shown at the top level of the application, but it is used on subsequent pages to go back to the previous page. The Search button allows access to the search form anywhere within the application. Here are the corresponding style definitions for each of these elements:

```
body > .toolbar {
    box-sizing: border-box;
    -webkit-box-sizing: border-box;
    -moz-box-sizing: border-box;
    border-bottom: 1px solid #2d3642;
    border-top: 1px solid #6d84a2;
    padding: 10px;
```

```css
    height: 45px;
    background: url(toolbar.png) #6d84a2 repeat-x;
}
.toolbar > h1 {
    position: absolute;
    overflow: hidden;
    left: 50%;
    margin: 1px 0 0 -75px;
    height: 45px;
    font-size: 20px;
    width: 150px;
    font-weight: bold;
    text-shadow: rgba(0, 0, 0, 0.4) 0px -1px 0;
    text-align: center;
    text-overflow: ellipsis;
    white-space: nowrap;
    color: #FFFFFF;
}
body[orient="landscape"] > .toolbar > h1 {
    margin-left: -125px;
    width: 250px;
}
.button {
    position: absolute;
    overflow: hidden;
    top: 8px;
    right: 6px;
    margin: 0;
    border-width: 0 5px;
    padding: 0 3px;
    width: auto;
    height: 30px;
    line-height: 30px;
    font-family: inherit;
    font-size: 12px;
    font-weight: bold;
    color: #FFFFFF;
    text-shadow: rgba(0, 0, 0, 0.6) 0px -1px 0;
    text-overflow: ellipsis;
    text-decoration: none;
    white-space: nowrap;
    background: none;
    -webkit-border-image: url(toolButton.png) 0 5 0 5;
}
#backButton {
    display: none;
    left: 6px;
    right: auto;
    padding: 0;
    max-width: 55px;
    border-width: 0 8px 0 14px;
    -webkit-border-image: url(backButton.png) 0 8 0 14;
}
```

The `body > .toolbar` class style is set to 45px in height. The `.toolbar > h1` header emulates the standard look of an application caption when in portrait mode, and `body[orient="landscape"] > .toolbar > h1` updates the position for landscape mode. Notice that the limited width of the iPhone viewport dictates use of `overflow:hidden` and `text-overflow:ellipsis`.

Adding a Top-Level Navigation Menu

Once the toolbar is created, you need to create the top-level navigation menu. Under the iUI framework, use a `ul` list, such as the following:

```
<ul id="home" title="iRealtor" selected="true">
    <li><a href="featured.html">Featured Listings</a></li>
    <li><a href="listings.html">All Listings</a></li>
    <li><a href="tips.html">Buying & Tips</a></li>
    <li><a href="calc.html">Mortgage Calculator</a></li>
    <li><a href="#meet_our_team">Meet Our Team</a></li>
    <li><a href="contact_us.html">Contact Us</a></li>
    <li><a href="index.html" target="_self">Visit our Web Site</a></li>
</ul>
```

iUI uses the `title` attribute to display in the toolbar's `h1` header. The `selected` attribute indicates that this `ul` element is the active block when the application loads. Each of the menu items is defined as an a link inside of `li` items. The `href` attribute can point to another `div` or `ul` block inside the same file (called a *panel*) using an anchor reference (such as #meet_our_team). Alternatively, you can use Ajax to load a block element from an external URL. Table 6-1 displays the four types of links you can work with inside iUI.

Table 6-1: iUI Link Types

Link Type	Description	Syntax
Internal URL	Loads a panel that is defined inside the same HTML page	`<a href="#meet_our_team">`
Ajax URL	Loads a document fragment via Ajax	`<a href="listings.html">`
Ajax URL Replace	Loads a document fragment via Ajax, replacing the contents of the calling link	`<a href="listings1.html" target="_replace">`
External URL	Loads an external Web link	`<a href="index.html" target="_self">`

The styles for the list items and links are as follows:

```
body > ul > li {
    position: relative;
    margin: 0;
    border-bottom: 1px solid #E0E0E0;
```

```css
        padding: 8px 0 8px 10px;
        font-size: 20px;
        font-weight: bold;
        list-style: none;
    }
body > ul > li > a {
    display: block;
    margin: -8px 0 -8px -10px;
    padding: 8px 32px 8px 10px;
    text-decoration: none;
    color: inherit;
    background: url(listArrow.png) no-repeat right center;
    }
```

Notice that `listArrow.png` is displayed at the right side of the list item's a link.

Displaying a Panel with an Internal URL

If you are linking to another block section inside the same page, you simply need to add the code. For example, the Meet Our Team item links to the following `div`:

```html
<div id="meet_our_team" class="panel" title="Meet Our Team">
    <h2>J-Team Realty</h2>
    <fieldset>
    <p class="normalText">Lorem ipsum dolor sit amet, consect etuer adipis
    cing elit. Suspend isse nisl. Vivamus a ligula vel quam tinci dunt posuere.
    Integer venen atis blandit est. Phasel lus ac neque. Quisque at augue.
    Phasellus purus. Sed et risus. Suspe ndisse laoreet consequat metus. Nam
    nec justo vitae tortor fermentum interdum. Aenean vitae quam eu urna
    pharetra ornare.</p>
    <p class="normalText">Pellent esque habitant morbi tristique senectus et
    netus et malesuada fames ac turpis egestas. Aliquam congue. Pel lentesque
    pretium fringilla quam. Integer libero libero, varius ut, faucibus et,
    facilisis vel, odio. Donec quis eros eu erat ullamc orper euismod. Nam
    aliquam turpis. Nunc convallis massa non sem. Donec non odio. Sed non lacus
    eget lacus hend rerit sodales.</p>
    </fieldset>
</div>
```

The `id` attribute value of the block element is identical to the `href` value of the source link (except for the # sign). The `div` element is assigned the `panel` class, and the `title` attribute supplies the new page title for the application. Inside the `div` element, the h2 element provides a header, whereas the `field-set` element, which is commonly used as a container inside iUI destination pages, houses the content. Figure 6-2 displays the results (based in part on additional styles that will be described shortly).

The `panel` class and `fieldset` styles are shown in the following code. In addition, the default h2 style is provided (although I will be updating this style in my own `irealtor.css` file):

```css
body > .panel {
    box-sizing: border-box;
    -webkit-box-sizing: border-box;
```

```css
        padding: 10px;
        background: #c8c8c8 url(pinstripes.png);
}
.panel > fieldset {
        position: relative;
        margin: 0 0 20px 0;
        padding: 0;
        background: #FFFFFF;
        -webkit-border-radius: 10px;
        border: 1px solid #999999;
        text-align: right;
        font-size: 16px;
}
.panel > h2 {
        margin: 0 0 8px 14px;
        font-size: inherit;
        font-weight: bold;
        color: #4d4d70;
        text-shadow: rgba(255, 255, 255, 0.75) 2px 2px 0;
}
```

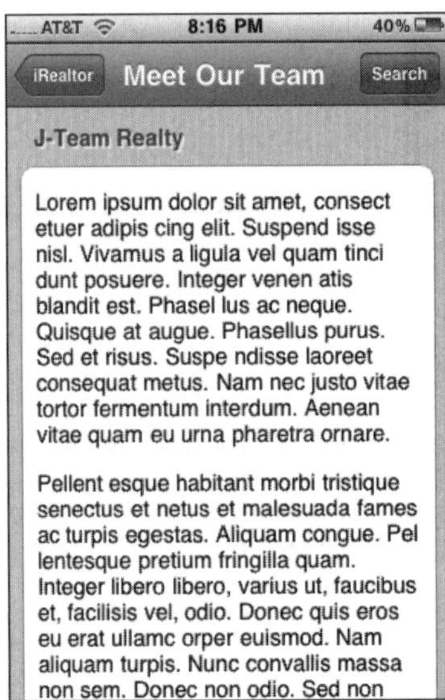

Figure 6-2: Destination page

The panel class property displays the vertical pinstripes, which is a standard background for iPhone and iPod touch applications. The fieldset, used primarily for displaying rows, is employed because it provides a white background box around the text content the page will display. However, because the iui.css styles did not display the margin/padding properties of h2 or p text as I needed it to, I linked irealtor.html with a new style sheet by placing the following declaration *below* the iui.css declaration:

```
<style type="text/css" media="screen">@import "irealtor.css";</style>
```

Inside of `irealtor.css`, the following styles are defined:

```css
.panel p.normalText {
    text-align: left;
    padding: 0 10px 0 10px;
}
.panel > h2 {
    margin: 3px 0 10px 10px;
}
```

Displaying Ajax Data from an External URL

You can create an entire iPhone Web application inside a single HTML page using internal links. However, this single-page approach breaks down when you begin to deal with large amounts of data. Therefore, you can use Ajax to break up your application into chunks, yet still maintain the same integrated look and feel of a single page app. When you use Ajax, iUI and other frameworks allow you to load content into your application on demand by providing an external URL. However, the document that is retrieved should be a document fragment, not a complete HTML page.

iUI fully encapsulates `XMLHttpRequest()` for you. Therefore, when you supply an external URL in a link that does not have `target="_self"` defined, it retrieves the document fragment and displays it.

In iRealtor, tapping the Featured Listings menu item (`<li><a href="featured.html">Featured Listings</a></li>`) should display a list of special homes that are being featured by this fictional local realtor. The contents of the file named `featured.html` are shown here:

```html
<ul id="featuredListings" title="Featured">
<li><a href="406509171.html">30 Bellview Ave, Bolton</a></li>
<li><a href="306488642.html">21 Milford Ave, Brandon</a></li>
<li><a href="326425649.html">10 Main St, Leominster</a></li>
<li><a href="786483624.html">12 Smuggle Lane, Marlboro</a></li>
<li><a href="756883629.html">34 Main Ave, Newbury</a></li>
<li><a href="786476262.html">33 Infinite Loop, Princeton</a></li>
<li><a href="706503711.html">233 Melville Road, Rutland</a></li>
<li><a href="767505714.html">320 Muffly, Sliver</a></li>
<li><a href="706489069.html">1 One Road, Zooly</a></li>
</ul>
```

The result is a basic navigation list, as shown in Figure 6-3. Each list item specifies a unique URL in which the app loads using Ajax when selected. You'll see this MLS listing destination page shortly.

The All Listings menu item illustrates some additional capabilities that you can add to a navigation list. Figure 6-4 displays the additional details added to the navigation list item, including a thumbnail picture and summary details in a second line.

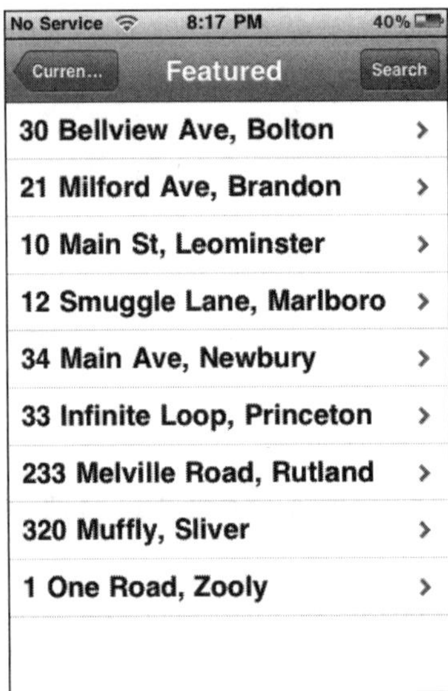

Figure 6-3: Listing data coming
from Ajax

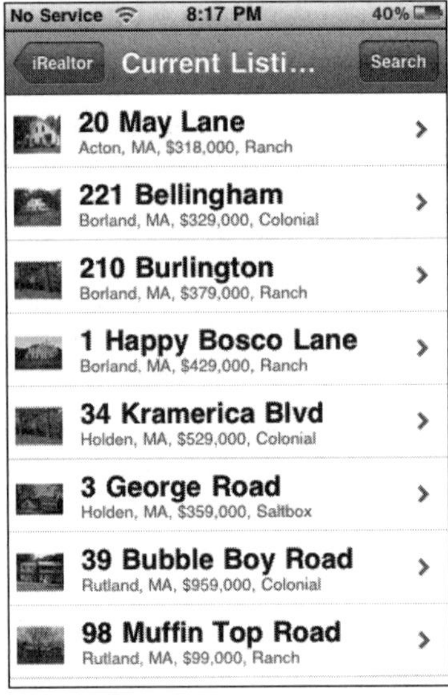

Figure 6-4: Enhanced navigational
menu items

The document fragment that is loaded via Ajax is as follows:

```
<ul id="listings" title="Current Listings">
<li>
```

```
        <img class="listingImg" src="images/406509171-sm.png" />
        <a class="listing" href="406509171.html">20 May Lane</a>
        <p class="listingDetails">Acton, MA, $318,000, Ranch</p>
    </li>
    <li>
        <img class="listingImg" src="images/306488642-sm.png" />
        <a class="listing" href="306488642.html">221 Bellingham</a>
        <p class="listingDetails">Borland, MA, $329,000, Colonial</p>
    </li>
    <li>
        <img class="listingImg" src="images/326425649-sm.png" />
        <a class="listing" href="326425649.html">210 Burlington</a>
        <p class="listingDetails">Borland, MA, $379,000, Ranch</p>
    </li>
    <li>
        <img class="listingImg" src="images/786483623-sm.png" />
        <a class="listing" href="786483624.html">1 Happy Bosco Lane</a>
        <p class="listingDetails">Borland, MA, $429,000, Ranch</p>
    </li>
    <li>
        <img class="listingImg" src="images/756883629-sm.png" />
        <a class="listing" href="756883629.html">34 Kramerica Blvd</a>
        <p class="listingDetails">Holden, MA, $529,000, Colonial</p>
    </li>
    <li>
        <img class="listingImg" src="images/786476262-sm.png" />
        <a class="listing" href="786476262.html">3 George Road</a>
        <p class="listingDetails">Holden, MA, $359,000, Saltbox</p>
    </li>
    <li>
        <img class="listingImg" src="images/706503711-sm.png" />
        <a class="listing" href="706503711.html">39 Bubble Boy Road</a>
        <p class="listingDetails">Rutland, MA, $959,000, Colonial</p>
    </li>
    <li>
        <img class="listingImg" src="images/767505713-sm.png" />
        <a class="listing" href="767505714.html">98 Muffin Top Road</a>
        <p class="listingDetails">Rutland, MA, $99,000, Ranch</p>
    </li>
    <li>
        <img class="listingImg" src="images/706489069-sm.png" />
        <a class="listing" href="706489069.html">1291 Blackjack Lane</a>
        <p class="listingDetails">Zambo, MA, $159,000, Saltbox</p>
    </li>
    </ul>
```

Each element inside the li element has a class style assigned to it. The following CSS styles are located in the irealtor.css file:

```
a.listing {
    padding-left: 54px;
```

```css
    padding-right: 40px;
    min-height: 34px;
}
img.listingImg {
    display: block;
    position: absolute;
    margin: 0;
    left: 6px;
    top: 7px;
    width: 35px;
    height: 27px;
    padding: 7px 0 10px 0;
}
p.listingDetails {
    display: block;
    position: absolute;
    margin: 0;
    left: 54px;
    top: 27px;
    text-align: left;
    font-size: 12px;
    font-weight: normal;
    color: #666666;
    text-decoration: none;
    width: 100%;
    height: 13px;
    padding: 3px 0 0 0;
}
```

The `img.listingImg` class positions the thumbnail at the far left side of the item. The
`p.listingDetails` class positions the summary text just below the main link.

Designing for Long Navigation Lists

Although a document fragment such as the one shown previously works fine for small amounts of
data, the performance would quickly drag with long lists. To deal with this issue, iUI allows you to
break large lists into manageable chunks by loading an initial set of items and then providing a link
to the next set (see Figure 6-5). This design emulates the way the iPhone Mail application works with
incoming messages.

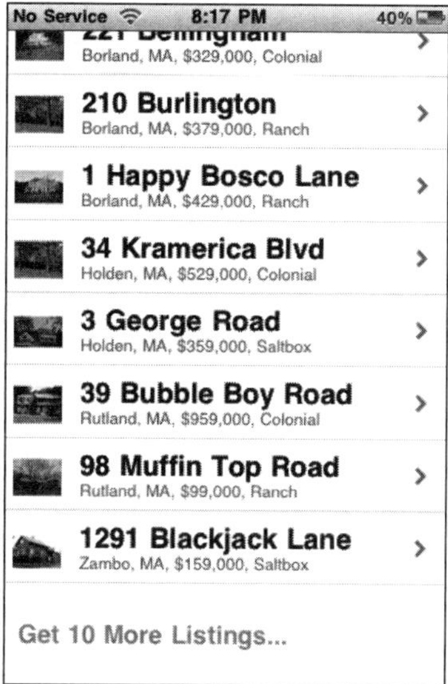

Figure 6-5: Loading additional listings

To provide this functionality in your application, create a link and add `target="_replace"` as an attribute. iUI loads the items from the URL, replacing the current link. As with other Ajax links, the URL needs to point to a document fragment, not a complete HTML file. Here's the link added to the bottom of the listings `ul` list:

```
<li><a href="listings1.html" target="_replace">Get 10 More Listings.</a></li>
```

When using the `target="_replace"` attribute, you need to use a fragment of a `ul` element and not a different structure. For example, the following document fragment is valid to use with a `_replace` request:

```
<li>item 1</li>
<li>item 2</li>
<li>item 3</li>
```

However, the following document fragment would not be correct because it is not valid inside a `ul` element:

```
<ul>
    <li>item 1</li>
    <li>item 2</li>
    <li>item 3</li>
</ul>
```

Creating a Destination Page

Each of the MLS listings in iRealtor has its own destination page that is accessed by an Ajax-based link, such as the following:

```
<a class="listing" href="406509171.html">20 May Lane</a>
```

The design goal of the page is to provide a picture and summary details of the house listing. But, taking advantage of iPhone's services, you also want to add a button for looking up the address in the Map app and an external Web link to a site providing town information. Figures 6-6 and 6-7 show the end design for this destination page.

The document fragment for this page is as follows:

```
<div title="20 May Lane" class="panel">
  <div>
    <img src="images/406509171.png" />
  </div>
  <h2>Details</h2>
  <fieldset>
      <div class="row">
        <label>mls #</label>
        <p>406509171</p>
      </div>
      <div class="row">
        <label>address</label>
        <p>20 May Lane</p>
      </div>
      <div class="row">
        <label>city</label>
        <p>Acton</p>
      </div>
      <div class="row">
        <label>price</label>
        <p>$318,000</p>
      </div>
      <div class="row">
        <label>type</label>
        <p>Single Family</p>
      </div>
      <div class="row">
        <label>acres</label>
        <p>0.27</p>
      </div>
      <div class="row">
        <label>rooms</label>
        <p>6</p>
      </div>
      <div class="row">
```

```
      <label>bath (f)</label>
      <p>1</p>
    </div>
    <div class="row">
      <label>bath (h)</label>
      <p>0</p>
    </div>
  </fieldset>
  <fieldset>
    <div class="row">
        <a  class="serviceButton" target="_self"
href="http://maps.google.com/maps?q=20+May+Lane,+Acton,+MA">Map To House</a>
    </div>
    <div class="row">
        <a  class="serviceButton" target="_self"
href="http://www.mass.gov/?pageID=mg2localgovccpage&L=3&L0=Home&L1=
State%20Government&L2=Local%20Government&sid=massgov2&selectCity=
Acton">View Town Info</a>
    </div>
  </fieldset>
</div>
```

There are several items of note. First, the `div` element is assigned the `panel` class, just as you did for the Meet Our Team page earlier in the chapter. Second, the individual items of the MLS listing data are contained in `div` elements with the `row` class. The set of `div row` elements is contained in a `fieldset`. Third, the button links to the map and external Web page are assigned a `serviceButton` class.

Figure 6-6: Top of listing page

Figure 6-7: Bottom of listing page

The styles for this page come from both `iui.css` and `irealtor.css`. First, here are the `row` class and `label` styles in `iui.css`. (If you recall, the `fieldset` is defined earlier in the chapter.)

```
.row  {
    position: relative;
    min-height: 42px;
    border-bottom: 1px solid #999999;
    -webkit-border-radius: 0;
    text-align: right;
}
fieldset > .row:last-child {
    border-bottom: none !important;
}
.row > label {
    position: absolute;
    margin: 0 0 0 14px;
    line-height: 42px;
    font-weight: bold;
}
```

The `row` class emulates the general look of an iPhone/iPod touch list row found in such locations as the built-in Settings and Contacts apps. The `.row:last-child` style removes the bottom border of the final row in a `fieldset`. The `.row > label` style defined in `iui.css` emulates the look of iPhone Settings, but as you will see in the following example, the code overrides this formatting to more closely emulate the Contacts look (right-aligned, black font).

The following styles are defined in `irealtor.css` to augment the base styles:

```css
.panel img {
    display: block;
    margin-left: auto;
    margin-right: auto;
    margin-bottom: 10px;
    border: 2px solid #666666;
    -webkit-border-radius: 6px;
}
.row > p {
    display: block;
    margin: 0;
    border: none;
    padding: 12px 10px 0 110px;
    text-align: left;
    font-weight: bold;
    text-decoration: inherit;
    height: 42px;
    color: inherit;
    box-sizing: border-box;
}
.row > label {
    text-align: right;
    width: 80px;
    position: absolute;
    margin: 0 0 0 14px;
    line-height: 42px;
    font-weight: bold;
    color: #7388a5;
}
.serviceButton {
    display: block;
    margin: 0;
    border: none;
    padding: 12px 10px 0 0px;
    text-align: center;
    font-weight: bold;
    text-decoration: inherit;
    height: 42px;
    color: #7388a5;
    box-sizing: border-box;
    -webkit-box-sizing: border-box;
}
```

The `.panel > img` centers the image with `margin-left:auto` and `margin-right:auto` and rounds the edges of the rectangle with `-webkit-border-radius`. (See Chapter 4, "Designing a Usable and Navigable UI," for more on this CSS style.)

The `.row > p` style formats the values of each MLS listing information. It is left-aligned and starts at 110px to the right of the left border of the element. The `.row > label` style adds specific formatting to emulate the Contacts UI look. The `.serviceButton` class style defines a link with a button look.

Adding a Dialog

The application pages that have been displayed have either been edge-to-edge navigation lists or destination panels for displaying content. However, you will probably need to create a modal dialog. When users are in a dialog, they need to either perform the intended action (such as a search or submittal) or cancel out. Just as in any desktop environment, a dialog is ideal for form entry.

iRealtor needs dialog boxes for two parts of the application: Search and Mortgage Calculator. The Search dialog is accessed by clicking the Search button on the top toolbar. Here's the calling link:

```
<a class="button" href="#searchForm">Search</a>
```

The link displays the internal link `#searchForm`. This references the `form` element with an `id` of `searchForm`:

```
<form id="searchForm" class="dialog" action="search.php">
    <fieldset>
        <h1>Search Listings</h1>
        <a class="button leftButton" type="cancel">Cancel</a>
        <a class="button blueButton" type="submit">Search</a>
        <select name="proptype" size="1">
          <option value="">Property Type</option>
          <option value="SF">Single-Family</option>
          <option value="CC">Condo</option>
          <option value="MF">Multi-Family</option>
          <option value="LD">Land</option>
          <option value="CI">Commercial</option>
          <option value="MM">Mobile Home</option>
          <option value="RN">Rental</option>
          <option value="BU">Business Opportunity</option>
        </select>
        <label class="altLabel">Min $:</label>
        <input class="altInput" type="text" name="minPrice" />
        <label class="altLabel">Max $:</label>
        <input class="altInput" type="text" name="maxPrice" />
        <label class="altLabel">MLS #:</label>
        <input class="altInput" type="text" name="mlsNumber" />
    </fieldset>
</form>
```

The `dialog` class indicates that the form is a dialog. The form elements are wrapped inside a `fieldset`. The action buttons for the dialog are actually defined as links. To be specific, the Cancel and Search links are defined as `button leftButton` and `button blueButton` classes, respectively. These two action buttons are displayed in the top toolbar of the dialog. It will also display the `h1` content as the dialog title.

A `select` list defines the type of properties that the user wants to choose from. Three `input` fields are defined for additional search criteria. Because the margin and padding styles are unique for this Search dialog, unique styles are specified for the `label` and `input` elements. You'll define those in a moment.

Figure 6-8 shows the form when displayed in the viewport. Per iPhone OS guidelines, the bottom part of the form is shaded to obscure the background page. Figure 6-9 displays the iPhone-specific selection list that is automatically displayed for you when the user clicks inside the `select` element. Finally, Figure 6-10 shows the pop-up keyboard that is displayed when the user clicks inside the `input` fields.

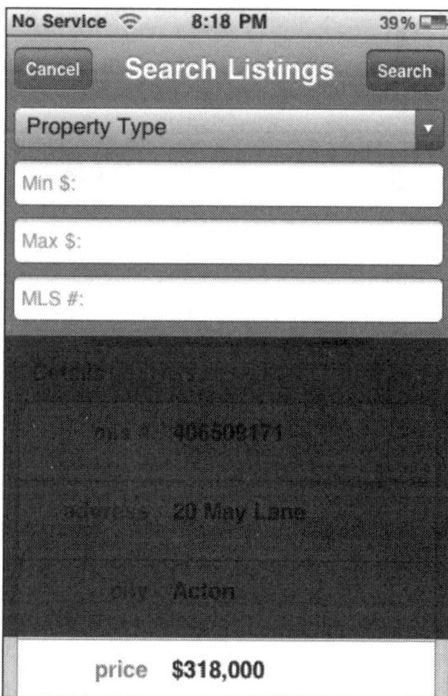

Figure 6-8: Search dialog box

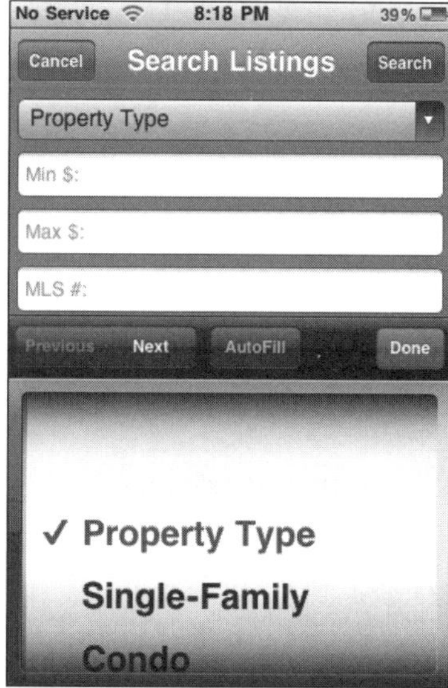

Figure 6-9: Select list items

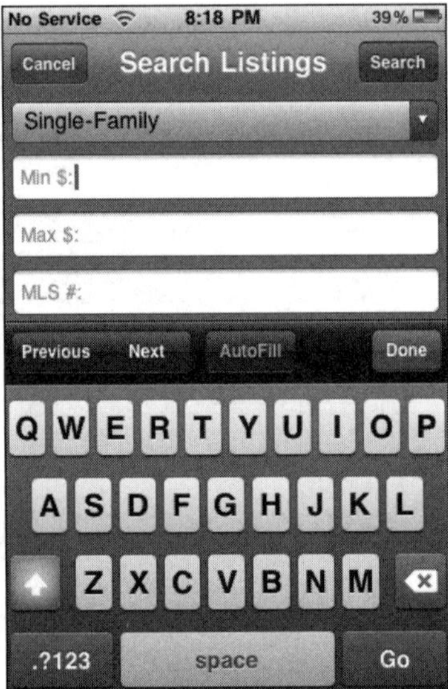

Figure 6-10: Text input of a form

Consider the CSS styles that are used to display this dialog. From the main stylesheet, there are several rules to pay attention to:

```css
body > .dialog {
    top: 0;
    width: 100%;
    min-height: 417px;
    z-index: 2;
    background: rgba(0, 0, 0, 0.8);
    padding: 0;
    text-align: right;
}
.dialog > fieldset {
    box-sizing: border-box;
    -webkit-box-sizing: border-box;
    width: 100%;
    margin: 0;
    border: none;
    border-top: 1px solid #6d84a2;
    padding: 10px 6px;
    background: url(toolbar.png) #7388a5 repeat-x;
}
.dialog > fieldset > h1 {
    margin: 0 10px 0 10px;
    padding: 0;
    font-size: 20px;
    font-weight: bold;
    color: #FFFFFF;
    text-shadow: rgba(0, 0, 0, 0.4) 0px -1px 0;
    text-align: center;
}
```

```css
.dialog > fieldset > label {
    position: absolute;
    margin: 16px 0 0 6px;
    font-size: 14px;
    color: #999999;
}
input {
    box-sizing: border-box;
    -webkit-box-sizing: border-box;
    width: 100%;
    margin: 8px 0 0 0;
    padding: 6px 6px 6px 44px;
    font-size: 16px;
    font-weight: normal;
}
.blueButton {
    -webkit-border-image: url(blueButton.png) 0 5 0 5;
    border-width: 0 5px;
}
.leftButton {
    left: 6px;
    right: auto;
}
```

The `body > .dialog` rule places the form over the entire application, including the top toolbar. It also defines a black background with 0.8 opacity. Notice the way in which the `.dialog > fieldset > label` style is defined so that the `label` element appears to be part of the `input` element. The `.blueButton` and `.leftButton` styles define the action button styles.

Three styles are defined in `irealtor.css` as an extension of `iui.css`:

```css
.altLabel {
    position: absolute;
    margin: 16px 15px 0 6px;
    font-size: 14px;
    color: black;
}
.altInput {
  padding-left: 60px;
}
select {
  box-sizing: border-box;
  -webkit-box-sizing: border-box;
  width: 100%;
  margin: 15px 0 0 0;
  padding: 6px 6px 6px 144px;
  font-size: 16px;
  font-weight: normal;
}
```

The `altLabel` and `altInput` rules appropriately size and position the `label` and `input` elements. The `select` rule styles the `select` element.

When you submit this form, it is submitted via Ajax to allow the results to slide from the side to provide a smooth transition.

You may, however, have other uses for dialogs beyond form submissions. For example, iRealtor includes a JavaScript-based mortgage calculator that is accessible from the top-level navigation menu. Here's the link:

```
<li><a href="calc.html">Mortgage Calculator</a></li>
```

The link accesses the document fragment contained in an external URL that contains the following form:

```
<form id="calculator" class="dialog">
  <fieldset>
      <h1>Mortgage Calculator</h1>
      <a class="button leftButton" type="cancel">Back</a>
      <label class="altLabel">Loan amount</label>
      <input class="calc" type="text" name="amt_zip" id="amt" />
      <label class="altLabel">Interest rate</label>
      <input class="calc" type="text" name="ir_zip" id="ir" />
      <label class="altLabel">Years</label>
      <input class="calc" type="text" name="amt_zip" id="term" onblur="calc()" />
      <label class="altLabel">Monthly payment</label>
      <input class="calc" type="text" readonly="true" id="payment" />
      <label class="altLabel">Total payment</label>
      <input class="calc" type="text" readonly="true" id="total" />
  </fieldset>
</form>
```

All the styles have been discussed already except for an additional one in `irealtor.css`:

```
input.calc {
    padding: 6px 6px 6px 120px;
}
```

This class style overrides the default padding to account for the longer labels used in the calculator.

The three `input` elements have a dummy `name` attribute that includes `zip`. The `zip` string prompts the numeric keyboard to display rather than the alphabet keyboard.

The purpose of the form is for the user to enter information into the first three `input` elements and then call the JavaScript function `calc()`, which then displays the results in the bottom two `input` fields. Because the calculation is performed inside a client-side JavaScript, no submittal is needed with the server.

The JavaScript function `calc()` needs to reside in the document head of the main `irealtor.html` file, not the document fragment. Here's the scripting code:

```
<script type="application/x-javascript">
function calc() {
    var amt = document.getElementById('amt').value;
    var ir =  document.getElementById('ir').value / 1200;
    var term =  document.getElementById('term').value * 12;
```

```javascript
   var total=1;
   for (var i = 0; i < term; i++) {
      total = total * (1 + ir);
   }
   var mp = amt * ir / ( 1 -- (1 / total));
   document.getElementById('payment').value = Math.round(mp * 100) / 100;
   document.getElementById('total').value = Math.round(mp * term * 100) / 100;
}
</script>
```

This routine performs a standard mortgage calculation and returns the results to the `payment` and `total` input fields. Figure 6-11 shows the result.

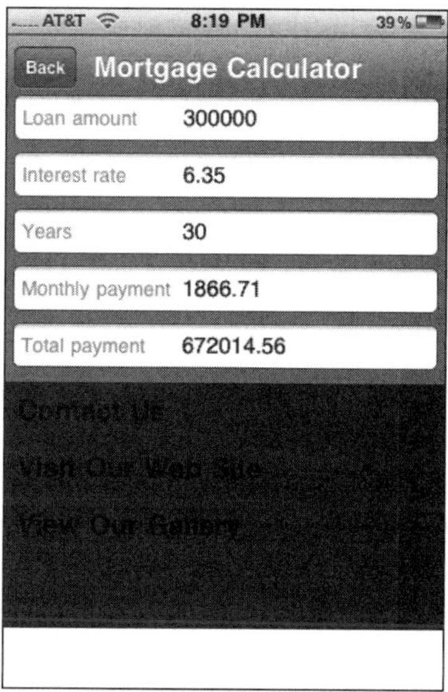

Figure 6-11: Text input of a form

Designing a Contact Us Page with Integrated iPhone Services

The final destination page of iRealtor is a Contact Us page that provides basic contact information for the local realtor and integrates with the Mail, Phone, and Map services of iPhone. The code is shown here.

The document fragment that is loaded by an Ajax external link is as follows:

```html
<div id="contact" title="Contact Us" class="panel">
   <div class="cuiHeader">
      <img class="cui" src="images/jordan_willmark.png" />
      <h1 class="cui" style="text-overflow:ellipsis;">Jordan Willmark</h1>
      <h2 class="cui">J-Team Realty</h2>
```

```
            </div>
            <fieldset>
                <div class="row">
                    <label class="cui">office</label>
                    <a class="cuiServiceLink" target="_self" href="tel:(978) 555-1212"
        onclick="return (navigator.userAgent.indexOf('iPhone') != -1)">(978) 555-1212</a>
                </div>
                <div class="row">
                    <label class="cui">mobile</label>
                    <a class="cuiServiceLink" target="_self" href="tel:(978) 545-1211"
        onclick="return (navigator.userAgent.indexOf('iPhone') != -1)">(978) 545-1211</a>
                </div>
                <div class="row">
                    <label class="cui">e-mail</label>
                    <a class="cuiServiceLink" target="_self"
        href="mailto:jordan@jteam3.com" onclick="return
        (navigator.userAgent.indexOf('iPhone') != -1)">jordan@jteam3.com</a>
                </div>
            </fieldset>
            <fieldset>
                <div class="rowCuiAddressBox">
                    <label class="cui">work</label>
                    <p class="cui">15 Louis Street</p>
                    <p class="cui">Princeton, MA 01541</p>
                </div>
            </fieldset>
            <fieldset>
                <div class="row">
                    <a   class="serviceButton" target="_self"
        href="http://maps.google.com/maps?q=15+Louis+St,+Princeton,+MA+(J-Team+Office)"
        >Map To Office</a>
                </div>
            </fieldset>
        </div>
```

You'll notice that the code listing displays several styles prefixed with `cui`. These are defined in a separate style sheet called `cui.css`. However, to use these styles, you need to add the following `style` element to the document head of `irealtor.html`:

```
<style type="text/css" media="screen">@import "./iui/cui.css";</style>
```

Figure 6-12 shows the panel when displayed on the iPhone.

The following three listings provide a full code view of the major source files that have been discussed. Listing 6-1 displays `irealtor.html`, Listing 6-2 provides `iui.css`, and Listing 6-3 contains `irealtor.css`.

Figure 6-12: iPhone-enabled Contact Us page

Listing 6-1: irealtor.html

```
<!DOCTYPE html PUBLIC "-//W3C//DTD XHTML 1.0 Strict//EN"
        "http://www.w3.org/TR/xhtml1/DTD/xhtml1-strict.dtd">
<html xmlns="http://www.w3.org/1999/xhtml">
<head>
<title>iRealtor</title>
<meta name="viewport" content="width=device-width; initial-scale=1.0;
maximum-scale=1.0; user-scalable=0;" />
<style type="text/css" media="screen">@import "./iui/iui.css";</style>
<style type="text/css" media="screen">@import "./iui/cui.css";</style>
<style type="text/css" media="screen">@import "irealtor.css";</style>
<script type="application/x-javascript" src="./iui/iui.js"></script>
<script type="application/x-javascript">
function calc() {
   var amt = document.getElementById('amt').value;
   var ir =  document.getElementById('ir').value / 1200;
   var term =  document.getElementById('term').value * 12;
   var total = 1;
   for (var i = 0; i < term; i++) {
      total = total * (1 + ir);
   }
   var mp = amt * ir / ( 1 -- (1 / total));
  document.getElementById('payment').value = Math.round(mp * 100) / 100;
```

Continued

Listing 6-1: irealtor.html *(continued)*

```
    document.getElementById('total').value = Math.round(mp * term *100) / 100;
}
</script>
</head>
<body>
    <!--Top toolbar-->
    <div class="toolbar">
        <h1 id="pageTitle"></h1>
        <a id="backButton" class="button" href="#"></a>
        <a class="button" href="#searchForm">Search</a>
    </div>
    <!--Home menu-->
    <ul id="home" title="iRealtor" selected="true">
        <li><a href="featured.html">Featured Listings</a></li>
        <li><a href="listings.html">All Listings</a></li>
        <li><a href="#">Buying & Tips</a></li>
        <li><a href="calc.html">Mortgage Calculator</a></li>
        <li><a href="#meet_our_team">Meet Our Team</a></li>
        <li><a href="contact_us.html">Contact Us</a></li>
        <li><a href="index.html" target="_self">Visit Our Web Site</a></li>
    </ul>
    <div id="meet_our_team" class="panel" title="Meet Our Team">
        <h2>J-Team Realty</h2>
        <fieldset>
        <p class="normalText">Lorem ipsum dolor sit amet, consect etuer adipis cing
elit. Suspend isse nisl. Vivamus a ligula vel quam tinci dunt posuere. Integer
venen atis blandit est. Phasel lus ac neque. Quisque at augue. Phasellus purus.
Sed et risus. Suspe ndisse laoreet consequat metus. Nam nec justo vitae tortor
fermentum interdum. Aenean vitae quam eu urna pharetra ornare.</p>
        <p class="normalText">Pellent esque habitant morbi tristique senectus et
netus et malesuada fames ac turpis egestas. Aliquam congue. Pel lentesque pretium
fringilla quam. Integer libero libero, varius ut, faucibus et, facilisis vel,
odio. Donec quis eros eu erat ullamc orper euismod. Nam aliquam turpis. Nunc
convallis massa non sem. Donec non odio. Sed non lacus eget lacus hend rerit
sodales.</p>
        </fieldset>
    </div>
    <form id="searchForm" class="dialog" action="search.php">
        <fieldset>
            <h1>Search Listings</h1>
            <a class="button leftButton" type="cancel">Cancel</a>
            <a class="button blueButton" type="submit">Search</a>
            <select name="proptype" size="1">
              <option value="">Property Type</option>
              <option value="SF">Single-Family</option>
              <option value="CC">Condo</option>
              <option value="MF">Multi-Family</option>
              <option value="LD">Land</option>
              <option value="CI">Commercial</option>
              <option value="MM">Mobile Home</option>
              <option value="RN">Rental</option>
              <option value="BU">Business Opportunity</option>
```

```html
                </select>
                <label class="altLabel">Min $:</label>
                <input type="text" name="minPrice" />
                <label class="altLabel">Max $:</label>
                <input type="text" name="maxPrice" />
                <label class="altLabel">MLS #:</label>
                <input type="text" name="mlsNumber" />
            </fieldset>
        </form>
    </body>
</html>
```

Listing 6-2: iui.css

```css
body {
    margin: 0;
    font-family: Helvetica;
    background: #FFFFFF;
    color: #000000;
    overflow-x: hidden;
    -webkit-user-select: none;
    -webkit-text-size-adjust: none;
}
body > *:not(.toolbar) {
    display: none;
    position: absolute;
    margin: 0;
    padding: 0;
    left: 0;
    top: 45px;
    width: 100%;
    min-height: 372px;
}
body[orient="landscape"] > *:not(.toolbar) {
    min-height: 268px;
}
body > *[selected="true"] {
    display: block;
}
a[selected], a:active {
    background-color: #194fdb !important;
    background-image: url(listArrowSel.png), url(selection.png)
!important;
    background-repeat: no-repeat, repeat-x;
    background-position: right center, left top;
    color: #FFFFFF !important;
}
a[selected="progress"] {
    background-image: url(loading.gif), url(selection.png)
!important;
}
/***************************************************************
*******************************/
```

Continued

Listing 6-2: iui.css *(continued)*

```css
body > .toolbar {
    box-sizing: border-box;
    -webkit-box-sizing: border-box;
    -moz-box-sizing: border-box;
    border-bottom: 1px solid #2d3642;
    border-top: 1px solid #6d84a2;
    padding: 10px;
    height: 45px;
    background: url(toolbar.png) #6d84a2 repeat-x;
}
.toolbar > h1 {
    position: absolute;
    overflow: hidden;
    left: 50%;
    margin: 1px 0 0 -75px;
    height: 45px;
    font-size: 20px;
    width: 150px;
    font-weight: bold;
    text-shadow: rgba(0, 0, 0, 0.4) 0px -1px 0;
    text-align: center;
    text-overflow: ellipsis;
    white-space: nowrap;
    color: #FFFFFF;
}
body[orient="landscape"] > .toolbar > h1 {
    margin-left: -125px;
    width: 250px;
}
.button {
    position: absolute;
    overflow: hidden;
    top: 8px;
    right: 6px;
    margin: 0;
    border-width: 0 5px;
    padding: 0 3px;
    width: auto;
    height: 30px;
    line-height: 30px;
    font-family: inherit;
    font-size: 12px;
    font-weight: bold;
    color: #FFFFFF;
    text-shadow: rgba(0, 0, 0, 0.6) 0px -1px 0;
    text-overflow: ellipsis;
    text-decoration: none;
    white-space: nowrap;
    background: none;
    -webkit-border-image: url(toolButton.png) 0 5 0 5;
}
  .blueButton {
```

```css
    -webkit-border-image: url(blueButton.png) 0 5 0 5;
    border-width: 0 5px;
}
.leftButton {
    left: 6px;
    right: auto;
}
#backButton {
    display: none;
    left: 6px;
    right: auto;
    padding: 0;
    max-width: 55px;
    border-width: 0 8px 0 14px;
    -webkit-border-image: url(backButton.png) 0 8 0 14;
}
.whiteButton,
.grayButton {
    display: block;
    border-width: 0 12px;
    padding: 10px;
    text-align: center;
    font-size: 20px;
    font-weight: bold;
    text-decoration: inherit;
    color: inherit;
}
.whiteButton {
    -webkit-border-image: url(whiteButton.png) 0 12 0 12;
    text-shadow: rgba(255, 255, 255, 0.7) 0 1px 0;
}
.grayButton {
    -webkit-border-image: url(grayButton.png) 0 12 0 12;
    color: #FFFFFF;
}
/*****************************************************************
*********************************/
body > ul > li {
    position: relative;
    margin: 0;
    border-bottom: 1px solid #E0E0E0;
    padding: 8px 0 8px 10px;
    font-size: 20px;
    font-weight: bold;
    list-style: none;
}
body > ul > li.group {
    position: relative;
    top: -1px;
    margin-bottom: -2px;
    border-top: 1px solid #7d7d7d;
    border-bottom: 1px solid #999999;
    padding: 1px 10px;
    background: url(listGroup.png) repeat-x;
```

Continued

Listing 6-2: iui.css *(continued)*

```css
        font-size: 17px;
        font-weight: bold;
        text-shadow: rgba(0, 0, 0, 0.4) 0 1px 0;
        color: #FFFFFF;
}
body > ul > li.group:first-child {
        top: 0;
        border-top: none;
}
body > ul > li > a {
        display: block;
        margin: -8px 0 -8px -10px;
        padding: 8px 32px 8px 10px;
        text-decoration: none;
        color: inherit;
        background: url(listArrow.png) no-repeat right center;
}
a[target="_replace"] {
        box-sizing: border-box;
        -webkit-box-sizing: border-box;
        padding-top: 25px;
        padding-bottom: 25px;
        font-size: 18px;
        color: cornflowerblue;
        background-color: #FFFFFF;
        background-image: none;
}
/************************************************************
********************************/
body > .dialog {
        top: 0;
        width: 100%;
        min-height: 417px;
        z-index: 2;
        background: rgba(0, 0, 0, 0.8);
        padding: 0;
        text-align: right;
}
.dialog > fieldset {
        box-sizing: border-box;
        -webkit-box-sizing: border-box;
        width: 100%;
        margin: 0;
        border: none;
        border-top: 1px solid #6d84a2;
        padding: 10px 6px;
        background: url(toolbar.png) #7388a5 repeat-x;
}
.dialog > fieldset > h1 {
        margin: 0 10px 0 10px;
        padding: 0;
        font-size: 20px;
        font-weight: bold;
```

```css
        color: #FFFFFF;
        text-shadow: rgba(0, 0, 0, 0.4) 0px -1px 0;
        text-align: center;
}
.dialog > fieldset > label {
        position: absolute;
        margin: 16px 0 0 6px;
        font-size: 14px;
        color: #999999;
}
input {
        box-sizing: border-box;
        -webkit-box-sizing: border-box;
        width: 100%;
        margin: 8px 0 0 0;
        padding: 6px 6px 6px 44px;
        font-size: 16px;
        font-weight: normal;
}
/*************************************************************
*********************************/
body > .panel {
        box-sizing: border-box;
        -webkit-box-sizing: border-box;
        padding: 10px;
        background: #c8c8c8 url(pinstripes.png);
}
.panel > fieldset {
        position: relative;
        margin: 0 0 20px 0;
        padding: 0;
        background: #FFFFFF;
        -webkit-border-radius: 10px;
        border: 1px solid #999999;
        text-align: right;
        font-size: 16px;
}
.row  {
        position: relative;
        min-height: 42px;
        border-bottom: 1px solid #999999;
        -webkit-border-radius: 0;
        text-align: right;
}
fieldset > .row:last-child {
        border-bottom: none !important;
}
.row > input {
        box-sizing: border-box;
        -webkit-box-sizing: border-box;
        margin: 0;
        border: none;
        padding: 12px 10px 0 110px;
        height: 42px;
        background: none;
```

Continued

Listing 6-2: iui.css *(continued)*

```css
}
.row > label {
    position: absolute;
    margin: 0 0 0 14px;
    line-height: 42px;
    font-weight: bold;
}
.row > .toggle {
    position: absolute;
    top: 6px;
    right: 6px;
    width: 100px;
    height: 28px;
}
.toggle {
    border: 1px solid #888888;
    -webkit-border-radius: 6px;
    background: #FFFFFF url(toggle.png) repeat-x;
    font-size: 19px;
    font-weight: bold;
    line-height: 30px;
}
.toggle[toggled="true"] {
    border: 1px solid #143fae;
    background: #194fdb url(toggleOn.png) repeat-x;
}
.toggleOn {
    display: none;
    position: absolute;
    width: 60px;
    text-align: center;
    left: 0;
    top: 0;
    color: #FFFFFF;
    text-shadow: rgba(0, 0, 0, 0.4) 0px -1px 0;
}
.toggleOff {
    position: absolute;
    width: 60px;
    text-align: center;
    right: 0;
    top: 0;
    color: #666666;
}
.toggle[toggled="true"] > .toggleOn {
    display: block;
}
.toggle[toggled="true"] > .toggleOff {
    display: none;
}
.thumb {
    position: absolute;
    top: -1px;
```

```css
        left: -1px;
        width: 40px;
        height: 28px;
        border: 1px solid #888888;
        -webkit-border-radius: 6px;
        background: #ffffff url(thumb.png) repeat-x;
}
.toggle[toggled="true"] > .thumb {
        left: auto;
        right: -1px;
}
.panel > h2 {
        margin: 0 0 8px 14px;
        font-size: inherit;
        font-weight: bold;
        color: #4d4d70;
        text-shadow: rgba(255, 255, 255, 0.75) 2px 2px 0;
}
/*****************************************************************
*********************************/
#preloader {
        display: none;
        background-image: url(loading.gif), url(selection.png),
            url(blueButton.png), url(listArrowSel.png), url(listGroup.png);
}
```

Listing 6-3: irealtor.css

```css
a.listing {
        padding-left: 54px;
        padding-right: 40px;
        min-height: 34px;
}
img.listingImg {
        display: block;
        position: absolute;
        margin: 0;
        left: 6px;
        top: 7px;
        width: 35px;
        height: 27px;
        padding: 7px 0 10px 0;
}
p.listingDetails {
        display: block;
        position: absolute;
        margin: 0;
        left: 54px;
        top: 27px;
        text-align: left;
        font-size: 12px;
        font-weight: normal;
        color: #666666;
        text-decoration: none;
```

Continued

Listing 6-3: irealtor.css *(continued)*

```css
    width: 100%;
    height: 13px;
    padding: 3px 0 0 0;
}
.panel img {
    display: block;
    margin-left: auto;
    margin-right: auto;
    margin-bottom: 10px;
     border: 2px solid #666666;
    -webkit-border-radius: 6px;
}
.row > p {
    display: block;
    margin: 0;
    border: none;
    padding: 12px 10px 0 110px;
     text-align: left;
    font-weight: bold;
    text-decoration: inherit;
    height: 42px;
    color: inherit;
    box-sizing: border-box;
    -webkit-box-sizing: border-box;
}
.row > label {
    text-align: right;
    width: 80px;
    position: absolute;
    margin: 0 0 0 14px;
    line-height: 42px;
    font-weight: bold;
    color: #7388a5;
}
.serviceButton {
    display: block;
    margin: 0;
    border: none;
    padding: 12px 10px 0 0px;
     text-align: center;
    font-weight: bold;
    text-decoration: inherit;
    height: 42px;
    color: #7388a5;
    box-sizing: border-box;
    -webkit-box-sizing: border-box;
}
/********************************/
.panel p.normalText {
    text-align: left;
    padding: 0 10px 0 10px;
}
```

```css
.panel > h2 {
    margin: 3px 0 10px 10px;
}
input.calc {
    padding: 6px 6px 6px 120px;
}
/********************************/
.altLabel {
  position: absolute;
  margin: 16px 15px 0 6px;
  font-size: 14px;
  color: black;
}
.altInput {
  padding-left: 60px;
}
select {
  box-sizing: border-box;
  -webkit-box-sizing: border-box;
  width: 100%;
  margin: 15px 0 0 0;
  padding: 6px 6px 6px 144px;
  font-size: 16px;
  font-weight: normal;
}
```

Scripting UI Behavior

When you use a framework such as iUI, you'll have a main JavaScript file (such as `iui.js`) that powers all the UI behavior for you once you include it in your document head. However, because the framework takes control over many aspects of the environment, it is important that you have a solid understanding of the library's internals.

I'll show you `iui.js` as an example. It consists of the object `window.iui`, three listeners for `load` and `click` events, and several supporting routines. All the JavaScript code is enclosed in an anonymous function with several constants and variables defined:

```javascript
(function() {
var slideSpeed = 20;
var slideInterval = 0;
var currentPage = null;
var currentDialog - null;
var currentWidth = 0;
var currentHash = location.hash;
var hashPrefix = "#_";
var pageHistory = [];
var newPageCount = 0;
var checkTimer;
// **** REST OF IUI CODE HERE ****
})();
```

The anonymous function creates a local scope to allow private semiglobal variables and avoid name conflicts with applications that use `iui.js`.

On Document Load

When the HTML document loads, the following listener function is triggered:

```javascript
addEventListener("load", function(event)
{
    var page = iui.getSelectedPage();
    if (page)
        iui.showPage(page);
    setTimeout(preloadImages, 0);
}, false);
```

The `getSelectedPage()` method of the JSON object `iui` is called to get the selected page — the block element node that contains a `selected="true"` attribute. This node is then passed to `iui.showPage()`, which is the core routine to display content.

`setTimeout()` is often used when calling certain JavaScript routines to prevent timing inconsistencies. Using `setTimeout()`, iUI calls an image preloader function to load application images.

Getting back to `iui.showPage()`, its code is as follows:

```javascript
showPage: function(page, backwards)
{
    if (page)
    {
        if (currentDialog)
        {
            currentDialog.removeAttribute("selected");
            currentDialog = null;
        }
        if (hasClass(page, "dialog"))
            showDialog(page);
        else
        {
            var fromPage = currentPage;
            currentPage = page;
            if (fromPage)
                setTimeout(slidePages, 0, fromPage, page, backwards);
            else
                updatePage(page, fromPage);
        }
    }
}
```

The `currentDialog` semi-global variable is evaluated to determine whether a dialog is already displayed. (`currentDialog` is set in the `showDialog()` function.) This variable would be `null` when the document initially loads because of the line `var currentDialog = null;`.

The node is then evaluated to determine whether it is a dialog (containing `class="dialog"` as an attribute) or a normal page. Although the opening page of an iPhone/iPod touch is often a normal page, you may want to have a login or initial search dialog.

Loading a Standard iUI Page

For normal pages, iUI assigns the value of `currentPage` to the variable `fromPage` and then reassigns `currentPage` to the `page` parameter. If `fromPage` is not `null` (that is, every page after the initial page), iUI performs a slide-in animation with a function called `slidePages()`. The `fromPage`, `page`, and `backwards` variables are passed to `slidePages()`.

However, because this is the first time running this routine (and `fromPage` will equal `null`), the `updatePage()` function is called:

```
function updatePage(page, fromPage)
{
    if (!page.id)
        page.id = "__" + (++newPageCount) + "__";
    location.href = currentHash = hashPrefix + page.id;
    pageHistory.push(page.id);
    var pageTitle = $("pageTitle");
    if (page.title)
        pageTitle.innerHTML = page.title;
    if (page.localName.toLowerCase() == "form" && !page.target)
        showForm(page);
    var backButton = $("backButton");
    if (backButton)
    {
        var prevPage = $(pageHistory[pageHistory.length-2]);
        if (prevPage && !page.getAttribute("hideBackButton"))
        {
            backButton.style.display = "inline";
            backButton.innerHTML = prevPage.title ? prevPage.title: "Back";
        }
        else
            backButton.style.display = "none";
    }
}
```

The `updatePage()` function is responsible for updating the `pageHistory` array, which is required for enabling the Mobile Safari Back button to work even in single-page applications. The value of the node's `title` attribute is then assigned to be the `innerHTML` of the top toolbar's `h1` `pageTitle`.

If the page name contains the string `form`, the `showForm()` function is called. Otherwise, the routine continues, looking to see if a `backButton` element is defined in the toolbar. If it is, the page history and button title are updated.

Subsequent pages will always bypass the direct call to `updatePage()` and use the `slidePages()` function instead. Here is the code:

```
function slidePages(fromPage, toPage, backwards)
{
    var axis = (backwards ? fromPage: toPage).getAttribute("axis");
    if (axis == "y")
        (backwards ? fromPage: toPage).style.top = "100%";
    else
```

```
            toPage.style.left = "100%";
    toPage.setAttribute("selected", "true");
    scrollTo(0, 1);
    clearInterval(checkTimer);
    var percent = 100;
    slide();
    var timer = setInterval(slide, slideInterval);
    function slide()
    {
        percent -= slideSpeed;
        if (percent <= 0)
        {
            percent = 0;
            if (!hasClass(toPage, "dialog"))
                fromPage.removeAttribute("selected");
            clearInterval(timer);
            checkTimer = setInterval(checkOrientAndLocation, 300);
            setTimeout(updatePage, 0, toPage, fromPage);
        }
        if (axis == "y")
        {
            backwards
                ? fromPage.style.top = (100-percent) + "%"
              : toPage.style.top = percent + "%";
        }
        else
        {
            fromPage.style.left = (backwards ? (100-percent): (percent-100)) + "%";
            toPage.style.left = (backwards ? -percent: percent) + "%";
        }
    }
}
```

The primary purpose of `slidePages()` is to emulate the standard iPhone/iPod touch slide animation effect when you move between pages. It achieves this by using JavaScript timer routines to incrementally update the `style.left` property of the `fromPage` and the `toPage`. The `updatePage()` function (discussed previously) is called inside a `setTimeout` routine.

Handling Link Clicks

Because most of the user interaction with an iPhone Web application is tapping the interface to navigate the application, the event listener for link clicks is, in many ways, the "mission control center" for `iui.jss`. Check out the code:

```
addEventListener("click", function(event)
{
    var link = findParent(event.target, "a");
    if (link)
    {
        function unselect() { link.removeAttribute("selected"); }
        if (link.href && link.hash && link.hash != "#")
        {
```

```
            link.setAttribute("selected", "true");
            iui.showPage($(link.hash.substr(1)));
            setTimeout(unselect, 500);
        }
        else if (link == $("backButton"))
            history.back();
        else if (link.getAttribute("type") == "submit")
            submitForm(findParent(link, "form"));
        else if (link.getAttribute("type") == "cancel")
            cancelDialog(findParent(link, "form"));
        else if (link.target == "_replace")
        {
            link.setAttribute("selected", "progress");
            iui.showPageByHref(link.href, null, null, link, unselect);
        }
        else if (!link.target)
        {
            link.setAttribute("selected", "progress");
            iui.showPageByHref(link.href, null, null, null, unselect);
        }
        else
            return;
        event.preventDefault();
    }
}, true);
```

This routine evaluates the type of link:

❑   If it is an internal URL, the page is passed to `iui.showPage()`.

❑   If the `backButton` is clicked, `history.back()` is triggered.

❑   Dialog forms typically contain a Submit and Cancel button. If a Submit button is clicked, `submitForm()` is called. If a Cancel button is clicked, `cancelDialog()` is called. (The `submitForm()` and `cancelDialog()` functions are discussed later in this chapter.)

❑   External URLs that have `target="_replace"` or that do not have target defined are Ajax links. Both of these call the `iui.showPageByHref()` method.

❑   If the link is none of these, it is external with a `target="_self"` attribute defined. The default iUI behavior is suspended and the link is treated as normal.

Handling Ajax Links

When a user clicks an Ajax link, the click event listener (shown previously) calls the `iui.showPageByHref()` method:

```
showPageByHref: function(href, args, method, replace, cb)
{
    var req = new XMLHttpRequest();
    req.onerror = function()
    {
        if (cb)
```

```
                    cb(false);
            };
            req.onreadystatechange = function()
            {
                if (req.readyState == 4)
                {
                    if (replace)
                        replaceElementWithSource(replace, req.responseText);
                    else
                    {
                        var frag = document.createElement("div");
                        frag.innerHTML = req.responseText;
                        iui.insertPages(frag.childNodes);
                    }
                    if (cb)
                        setTimeout(cb, 1000, true);
                }
            };
            if (args)
            {
                req.open(method || "GET", href, true);
                req.setRequestHeader("Content-Type",
                "application/x-www-form-urlencoded");
                req.setRequestHeader("Content-Length", args.length);
                req.send(args.join("&"));
            }
            else
            {
                req.open(method || "GET", href, true);
                req.send(null);
            }
        }
```

The routine calls XMLHttpRequest() to assign the req object. If the args parameter is not null (that is, when an Ajax form is submitted), the form data is sent to the server. If args is null, the supplied URL is sent to the server. The processing of incoming text takes place inside the onreadystatechange handler, which handles the response from the server.

If replace is true (meaning that target="_replace" is specified in the calling link), the replaceElementWithSource() function is called. As the following code shows, the calling link node (the replace parameter) is replaced with the source (the Ajax document fragment):

```
function replaceElementWithSource(replace, source)
{
    var page = replace.parentNode;
    var parent = replace;
    while (page.parentNode != document.body)
    {
        page = page.parentNode;
        parent = parent.parentNode;
    }
    var frag = document.createElement(parent.localName);
    frag.innerHTML = source;
```

```
            page.removeChild(parent);
        while (frag.firstChild)
            page.appendChild(frag.firstChild);
    }
```

If a click is generated from a normal Ajax link, the contents of the external URL are displayed in a new page. Therefore, a `div` is created and the document fragment is added as the `innerHTML` of the element. The `iui.insertPages()` method adds the new nodes to create a new page, and this page is passed to `iui.showPage()`:

```
    insertPages: function(nodes)
        {
            var targetPage;
            for (var i = 0; i < nodes.length; ++i)
            {
                var child = nodes[i];
                if (child.nodeType == 1)
                {
                    if (!child.id)
                        child.id = "__" + (++newPageCount) + "__";
                    var clone = $(child.id);
                    if (clone)
                        clone.parentNode.replaceChild(child, clone);
                    else
                        document.body.appendChild(child);
                    if (child.getAttribute("selected") == "true" || !targetPage)
                        targetPage = child;
                }
            }
            if (targetPage)
                iui.showPage(targetPage);
        }
```

Loading a Dialog

If the node that is passed into the main `showPage()` function is a dialog (`class="dialog"`), the `show-Dialog()` function is called, which in turn calls `showForm()`. These two functions are shown in the following code:

```
    function showDialog(page)
    {
        currentDialog = page;
        page.setAttribute("selected", "true");
        if (hasClass(page, "dialog") && !page.target)
            showForm(page);
    }
    function showForm(form)
    {
        form.onsubmit = function(event)
        {
            event.preventDefault();
            submitForm(form);
        };
```

```
form.onclick = function(event)
{
    if (event.target == form && hasClass(form, "dialog"))
        cancelDialog(form);
};
}
```

The `showForm()` function assigns event handlers to the `onsubmit` and `onclick` events of the form. When a form is submitted, the `submitForm()` function submits the form data via Ajax. When an element on the form is clicked, the dialog is closed. The following code shows the routines that are called:

```
function submitForm(form)
{
    iui.showPageByHref(form.action || "POST", encodeForm(form), form.method);
}
function cancelDialog(form)
{
    form.removeAttribute("selected");
}
function encodeForm(form)
{
    function encode(inputs)
    {
        for (var i = 0; i < inputs.length; ++i)
        {
            if (inputs[i].name)
                args.push(inputs[i].name + "=" + escape(inputs[i].value));
        }
    }
    var args = [];
    encode(form.getElementsByTagName("input"));
    encode(form.getElementsByTagName("select"));
    return args;
}
```

The entire code for `iui.js` is provided in Listing 6-4.

Listing 6-4: iui.js

```
(function() {
var slideSpeed = 20;
var slideInterval = 0;
var currentPage = null;
var currentDialog = null;
var currentWidth = 0;
var currentHash = location.hash;
var hashPrefix = "#_";
var pageHistory = [];
var newPageCount = 0;
var checkTimer;
//
// *****************************************************************
// *******************************************
```

```
window.iui =
{
    showPage: function(page, backwards)
    {
        if (page)
        {
            if (currentDialog)
            {
                currentDialog.removeAttribute("selected");
                currentDialog = null;
            }
            if (hasClass(page, "dialog"))
                showDialog(page);
            else
            {
                var fromPage = currentPage;
                currentPage = page;
                if (fromPage)
                    setTimeout(slidePages, 0, fromPage, page,
                    backwards);
                else
                    updatePage(page, fromPage);
            }
        }
    },
    showPageById: function(pageId)
    {
        var page = $(pageId);
        if (page)
        {
            var index = pageHistory.indexOf(pageId);
            var backwards = index != -1;
            if (backwards)
                pageHistory.splice(index, pageHistory.length);
            iui.showPage(page, backwards);
        }
    },
    showPageByHref: function(href, args, method, replace, cb)
    {
        var req = new XMLHttpRequest();
        req.onerror = function()
        {
            if (cb)
                cb(false);
        };
        req.onreadystatechange = function()
        {
            if (req.readyState == 4)
            {
                if (replace)
                    replaceElementWithSource(replace, req.responseText);
                else
                {
                    var frag = document.createElement("div");
```

Listing 6-4: iui.js *(continued)*

```javascript
                    frag.innerHTML = req.responseText;
                    iui.insertPages(frag.childNodes);
                }
                if (cb)
                    setTimeout(cb, 1000, true);
            }
        };
        if (args)
        {
            req.open(method || "GET", href, true);
            req.setRequestHeader("Content-Type",
            "application/x-www-form-urlencoded");
            req.setRequestHeader("Content-Length", args.length);
            req.send(args.join("&"));
        }
        else
        {
            req.open(method || "GET", href, true);
            req.send(null);
        }
    },
    insertPages: function(nodes)
    {
        var targetPage;
        for (var i = 0; i < nodes.length; ++i)
        {
            var child = nodes[i];
            if (child.nodeType == 1)
            {
                if (!child.id)
                    child.id = "__" + (++newPageCount) + "__";
                var clone = $(child.id);
                if (clone)
                    clone.parentNode.replaceChild(child, clone);
                else
                    document.body.appendChild(child);
                if (child.getAttribute("selected") == "true"
                || !targetPage)
                    targetPage = child;
              --i;
            }
        }
        if (targetPage)
            iui.showPage(targetPage);
    },
    getSelectedPage: function()
    {
        for (var child = document.body.firstChild; child; child =
        child.nextSibling)
        {
            if (child.nodeType == 1 &&
            child.getAttribute("selected") == "true")
```

```
                        return child;
            }
        }
};
//
******************************************************************
**********************************
addEventListener("load", function(event)
{
    var page = iui.getSelectedPage();
    if (page)
        iui.showPage(page);
    setTimeout(preloadImages, 0);
    setTimeout(checkOrientAndLocation, 0);
    checkTimer = setInterval(checkOrientAndLocation, 300);
}, false);
addEventListener("click", function(event)
{
    var link = findParent(event.target, "a");
    if (link)
    {
        function unselect() { link.removeAttribute("selected"); }
        if (link.href && link.hash && link.hash != "#")
        {
            link.setAttribute("selected", "true");
            iui.showPage($(link.hash.substr(1)));
            setTimeout(unselect, 500);
        }
        else if (link == $("backButton"))
            history.back();
        else if (link.getAttribute("type") == "submit")
            submitForm(findParent(link, "form"));
        else if (link.getAttribute("type") == "cancel")
            cancelDialog(findParent(link, "form"));
        else if (link.target == "_replace")
        {
            link.setAttribute("selected", "progress");
            iui.showPageByHref(link.href, null, null, link,
            unselect);
        }
        else if (!link.target)
        {
            link.setAttribute("selected", "progress");
            iui.showPageByHref(link.href, null, null, null,
            unselect);
        }
        else
            return;
        event.preventDefault();
    }
}, true);
addEventListener("click", function(event)
{
    var div = findParent(event.target, "div");
```

Continued

Listing 6-4: iui.js *(continued)*

```javascript
        if (div && hasClass(div, "toggle"))
        {
            div.setAttribute("toggled", div.getAttribute("toggled")
            != "true");
            event.preventDefault();
        }
    }, true);
    function checkOrientAndLocation()
    {
        if (window.innerWidth != currentWidth)
        {
            currentWidth = window.innerWidth;
            var orient = currentWidth == 320 ? "profile": "landscape";
            document.body.setAttribute("orient", orient);
            setTimeout(scrollTo, 100, 0, 1);
        }
        if (location.hash != currentHash)
        {
            var pageId = location.hash.substr(hashPrefix.length)
            iui.showPageById(pageId);
        }
    }
    function showDialog(page)
    {
        currentDialog = page;
        page.setAttribute("selected", "true");
        if (hasClass(page, "dialog") && !page.target)
            showForm(page);
    }
    function showForm(form)
    {
        form.onsubmit = function(event)
        {
            event.preventDefault();
            submitForm(form);
        };
        form.onclick = function(event)
        {
            if (event.target == form && hasClass(form, "dialog"))
                cancelDialog(form);
        };
    }
    function cancelDialog(form)
    {
        form.removeAttribute("selected");
    }
    function updatePage(page, fromPage)
    {
        if (!page.id)
            page.id = "__" + (++newPageCount) + "__";
        location.href = currentHash = hashPrefix + page.id;
        pageHistory.push(page.id);
```

```
    var pageTitle = $("pageTitle");
    if (page.title)
        pageTitle.innerHTML = page.title;
    if (page.localName.toLowerCase() == "form" && !page.target)
        showForm(page);
    var backButton = $("backButton");
    if (backButton)
    {
        var prevPage = $(pageHistory[pageHistory.length-2]);
        if (prevPage && !page.getAttribute("hideBackButton"))
        {
            backButton.style.display = "inline";
            backButton.innerHTML = prevPage.title ?
            prevPage.title: "Back";
        }
        else
            backButton.style.display = "none";
    }
}
function slidePages(fromPage, toPage, backwards)
{
    var axis = (backwards ? fromPage: toPage).getAttribute("axis");
    if (axis == "y")
        (backwards ? fromPage: toPage).style.top = "100%";
    else
        toPage.style.left = "100%";
    toPage.setAttribute("selected", "true");
    scrollTo(0, 1);
    clearInterval(checkTimer);
    var percent = 100;
    slide();
    var timer = setInterval(slide, slideInterval);
    function slide()
    {
        percent -= slideSpeed;
        if (percent <= 0)
        {
            percent = 0;
            if (!hasClass(toPage, "dialog"))
                fromPage.removeAttribute("selected");
            clearInterval(timer);
            checkTimer = setInterval(checkOrientAndLocation, 300);
            setTimeout(updatePage, 0, toPage, fromPage);
        }
        if (axis -- "y")
        {
            backwards
                ? fromPage.style.top = (100-percent) + "%"
              : toPage.style.top = percent + "%";
        }
        else
        {
            fromPage.style.left = (backwards ? (100-percent):
            (percent-100)) + "%";
```

Continued

Listing 6-4: iui.js *(continued)*

```javascript
                toPage.style.left = (backwards ? -percent: percent)
                + "%";
            }
        }
    }
    function preloadImages()
    {
        var preloader = document.createElement("div");
        preloader.id = "preloader";
        document.body.appendChild(preloader);
    }
    function submitForm(form)
    {
        iui.showPageByHref(form.action || "POST", encodeForm(form),
        form.method);
    }
    function encodeForm(form)
    {
        function encode(inputs)
        {
            for (var i = 0; i < inputs.length; ++i)
            {
                if (inputs[i].name)
                    args.push(inputs[i].name + "=" +
                    escape(inputs[i].value));
            }
        }
        var args = [];
        encode(form.getElementsByTagName("input"));
        encode(form.getElementsByTagName("select"));
        return args;
    }
    function findParent(node, localName)
    {
        while (node && (node.nodeType != 1 ||
        node.localName.toLowerCase() != localName))
            node = node.parentNode;
        return node;
    }
    function hasClass(self, name)
    {
        var re = new RegExp("(^|\\s)"+name+"($|\\s)");
        return re.exec(self.getAttribute("class")) != null;
    }
    function replaceElementWithSource(replace, source)
    {
        var page = replace.parentNode;
        var parent = replace;
        while (page.parentNode != document.body)
        {
            page = page.parentNode;
            parent = parent.parentNode;
```

```
        }
        var frag = document.createElement(parent.localName);
        frag.innerHTML = source;
        page.removeChild(parent);
        while (frag.firstChild)
            page.appendChild(frag.firstChild);
    }
    function $(id) { return document.getElementById(id); }
    function ddd() { console.log.apply(console, arguments); }
})();
```

Summary

The focus of this chapter was to "roll up your sleeves" and dive into the nitty gritty programming details of a typical iPhone Web app user interface. Using iUI, I walked through the development of a fictional app called iRealtor. I explained how to create a side-to-side navigation page, a detail page, a contact form, as well as how to work with several different types of controls found on these pages.

7

Handling Touch Interactions and Events

An essential part of any Web application is the ability to respond to events triggered by the user or by a condition that occurs on the client: the clicking of a button, the pressing of a key, the scrolling of a window. Whereas the user interacts with an HTML element, the entire document, or the browser window, JavaScript serves as the watchful eye behind the scenes that monitors all this activity taking place and fires off events as they occur.

With its touch interface, iPhone is all about direct interactivity with the user. As a result, it is not surprising that any iPhone Web app you create can handle the variety of *gestures* — finger taps, flicks, swipes, and pinches — that a user naturally performs as they interact with your app on their mobile device.

The Three Types of Touch Events

There are three primary types of touch-related events to consider when developing an iPhone Web app. These include the following:

- ❑ **Mouse emulation events:** Events in which a one- or two-finger action simulates a mouse.

- ❑ **Touch events:** Events that dispatch when one or more fingers touch the screen. You can trap touch events and work with multiple finger touches that occur at different points on the screen at the same time.

- ❑ **Gesture events:** Combinations of touch events that also support scaling and rotation information.

The mouse emulation events are the ones that you'll be working with for most purposes inside your Web app. As such, I'll spend most of my time exploring them. However, I also talk about the touch and gesture events later in the chapter.

Mouse-Emulation Events

When working with touch interactions and events for iPhone, keep in mind that several of the gestures that a user performs are designed to emulate mouse events. At the same time, I need to make one thing clear: *the finger is not the same as a mouse.* As a result, the traditional event model that Web developers are so used to working with on desktop computers does not always apply as they may expect in this new context. The new ground rules are described in the following sections.

Many Events Are Handled by Default

By default, many of the events are handled for you automatically by iPhone and Safari. As a result, you don't need to write code to handle the basic touch interactions of the user. Flick-scrolling, zoom pinching and unpinching, and one-finger panning (or scrolling) are user inputs that come free. You can, however, trap for many of these events with touch or gesture events.

Conditional Events

The way in which Safari events are handled depends on two key variables:

- **Number of fingers:** Different events fire depending on whether a one-finger or two-finger gesture is performed. Tables 7-1 and 7-2 list the common iPhone one- and two-finger gestures and their ability to trap these events.

- **Event target:** Events are handled differently depending on the HTML element being selected. In particular, the event flow varies depending on whether the target element is clickable or scrollable. A clickable target is one that supports standard mouse events (for example, `mousemove`, `mousedown`, `mouseup`, `click`), whereas a scrollable target is one equipped to deal with overflow, scrollbars, and so on. (However, as I'll discuss later in this chapter, you can make any element clickable by registering an event handler for it.)

Table 7-1: One-Finger Gestures

Gesture	Event(s) Fired	Description
Panning	`onscroll` fired when gesture ends	No events triggered until user stops panning motion.
Touch and hold	None	Some native iPhone apps support this gesture. For example, touch and hold on an image displays a Save dialog box.
Double-tap	None	
Tap	*Clickable element:* `mouseover`, `mousemove`, `mousedown`, `mouseup`, `click` *Nonclickable element:* None	If `mouseover` or `mousemove` changes page content, remaining events in sequence are cancelled.

Table 7-2: Two-Finger Gestures

Gesture	Event(s) Fired	Description
Pinch/unpinch zoom	None	Gesture used to zoom and unzoom on a Web page or image.
Two-finger panning	*Scrollable element:* `mouseevent` *Nonscrollable element:* `None`	If nonscrollable element, event is treated as a page-level scroll (which fires an `onscroll` when gesture stops).

Mouse Events: Think "Click," Not "Move"

The general rule of thumb for iPhone event handling is that no events trigger *until* the user's finger leaves the touch screen. As a result, the normal flow of mouse events is radically altered over the traditional browser event model, because mouse events now depend on the actual selection (or clicking) of an element, not simply when a finger passes over it. Said differently, the focus or selection of an element is what matters, not the movement of the mouse.

Take the example of a page that consists of two links: Link A and Link B. Suppose a user glides his finger from Link A, passes over the top of Link B, and then clicks on it. Here's how the events will be handled:

- ❑ The `mouseover`, `mousemove`, and `mousedown` event handlers of Link B are fired only *after* a `mouseup` event occurs (but before `mouseup` is triggered). As a result, from a practical standpoint, these preliminary mouse events are rendered useless.

- ❑ Because you can't perform a `mousedown` and `mouseup` without a `click`, they all refer to the same event.

- ❑ The `mouseout` event of Link A is fired only after the user clicks on Link B, not when the finger moves off of Link A. The CSS pseudo-style `:hover` is applied to Link B only when the user selects it and is removed from Link A only when Link B is actually selected, not before.

Therefore, for most purposes, you should design your app to respond to click events rather than other mouse events.

Click-Enabling Elements

If the element you are working with is not considered "clickable" by Safari, all mouse-related events for it are ignored. However, this can be problematic in certain situations, such as if you are using `span` or `div` elements in a cascading menu and want to change the menu display based on the context of the user.

To override Safari's default behavior, you need to force the element to be considered "clickable" by Safari. To do so, you assign an empty `click` handler to the element (`click="void(0)"`). For example:

```
<span mousemove="displayMenu(event)" click="void(0)">Run Lola Run</span>
```

Once you add the empty `click` handler, the other mouse-related events begin to fire.

Event Flow

Besides the anomaly of the timing of the `mousedown` event, the rest of the supported mouse and key events fire in Safari in the same sequence as a standard Web browser. Table 7-3 shows the event sequences that occur when both a block-level element and a form element are clicked. The form element column also displays the order of key events if the user types in the on-screen keyboard.

Table 7-3: Event Sequencing

Clickable Block-Level Elements (e.g., Link)	Form Element (e.g., Textarea, Input)
mouseover	mouseover
mousedown	mousedown
mouseup	focus
click	mouseup

mouseout (only after the next element receives focus)	click
	    keydown
	    keypress
	    keyup
	    change
	
	blur (only after the next element receives focus)
	
	mouseout (only after the next element receives focus)

Unsupported Events

You cannot trap for all events inside Safari. For example, you cannot trap for events associated with a user switching between pages in Safari. The `focus` and `blur` events of the `window` object are not triggered when the focus moves off or on a page. Additionally, when another page becomes the active page, JavaScript events (including polling events created with `setInterval()`) are not fired. However, the `unload` event of the `window` object is triggered when the user loads a new page in the current window.

Table 7-4 lists the events that are fully supported and unsupported.

Table 7-4: Event Compatibility

Supported Events	Unsupported Events
click	cut
mouseout*	copy
mouseover*	paste

Supported Events	Unsupported Events
mouseup*	drag
mousedown*	drop
mouseout*	dblclick
blur	selection
change	formfield.onmouseenter
focus	formfield.onmouseleave
load	formfield.onmousemove
unload	formfield.onselect
reset	contextmenu
mousewheel	error
submit	resize
abort	scroll
orientationchange	
touchstart	
touchmove	
Touchend	
Touchcancel	
Gesturestart	
Gesturechange	
Gestureend	

* Different behavior than in a desktop browser. (See the "Mouse Events: Think 'Click,' Not 'Move'" section earlier in this chapter.)

Capturing Two-Finger Scrolling

Using mouse-emulation events, you can program the two-finger scroll. Whereas a one-finger scroll is used to move an entire page around, the two-finger scroll can be used to scroll inside any scrollable region of a page, such as a text area. Because Safari on iPhone supports the overriding of the `window.mousewheel` event, you can use the two-finger scroll for your own purposes.

Suppose, for example, that you want to control the vertical position of a ball image based on the two-finger scroll input of the user inside a scrollable region. When the user scrolls up, you want the ball to move up. When the user scrolls down, you want the ball to move down. Figure 7-1 shows the UI layout for this example.

Start with the page layout and styles:

```
<!DOCTYPE html PUBLIC "-//W3C//DTD XHTML 1.0 Strict//EN"
        "http://www.w3.org/TR/xhtml1/DTD/xhtml1-strict.dtd">
<html xmlns="http://www.w3.org/1999/xhtml">
<head>
<title>ScrollPad</title>
<meta name="viewport" content="width=320; initial-scale=1.0; maximum-scale=1.0;
user-scalable=0;">
<style type="text/css" media="screen">
    body {
        margin: 0;
        padding: 0;
        width: 320px;
      height: 416px;
        font-family: Helvetica;
        -webkit-user-select: none;
        cursor: default;
        -webkit-text-size-adjust: none;
    background: #000000;
    color: #FFFFFF;
    }
    #leftPane {
        position: absolute;
        width: 160px;
        height: 100%;
    }
    #rightPane {
        position: absolute;
        width: 140px;
        left: 161px;
        height:100%;
    }
  #scrollPad {
        width: 148px;
        top: 3px;
        height: 300px;
        border-style: none;
        background-image: url( 'fs.png' );
    }
    #blueDot {
            position: absolute;
            left: 50px;
            top: 10px;
    }
</style>
</head>
<body>
    <div id="leftPane">
        <p>Use a two-finger scroll in the scrollpad to move the blue dot.</p>
        <form>
        <textarea id="scrollPad" readonly="readonly" disabled="true"></textarea>
        </form>
    </div>
```

```
    <div id="rightPane">
        <img id="blueDot" src="compose_atom_selected.png"/>
    </div>
</body>
</html>
```

The `scrollPad textarea` element is used as the *hot* scrollable region. It is enclosed inside a `div` on the left half of the page and sized large enough so that a two-finger scroll is easy for people to perform inside its borders. To ensure that the `textarea` is easy to identify on the screen, an arrow PNG is added as the background image and a solid border is defined. The `disabled="true"` attribute value must be added to prevent keyboard input in the control. On the other side of the page, the `blueDot img` is enclosed inside a `div` on the right.

The interactivity comes by capturing `window.mousewheel`, which is the event that Safari triggers when a user performs a two-finger scroll. You do that through an `addEventListener()` call that you add to a `<script>` tag in the document head.

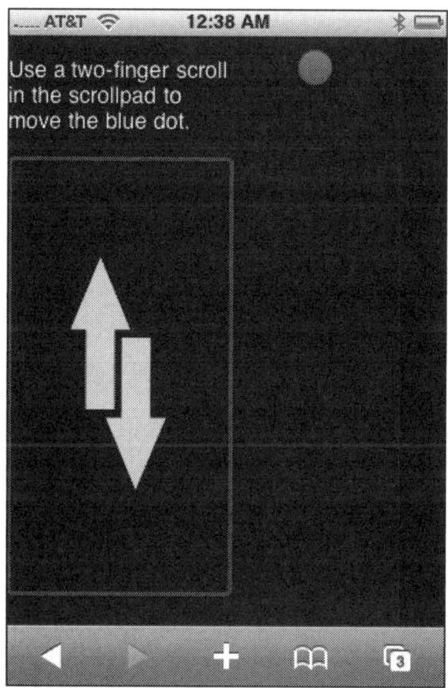

Figure 7-1: UI for the ScrollPad application

```
<script type="application/x-javascript">
    addEventListener('load', function() {window.mousewheel = twoFingerScroll; });
</script>
```

As shown in the preceding example, a function called `twoFingerScroll()` is assigned to be the event handler for `window.mousewheel`.

Next, here's the code for `twoFingerScroll()`:

```
function twoFingerScroll(wEvent)
{
    var delta = wEvent.wheelDelta / 120;
```

```
        scrollBall(delta);
        return true;
}
```

You can place this function and all of the other code in the same <script> block in the document head.

The `wheelDelta` property returns negative 120 when the scroll movement is upward and positive 120 when the movement is downward. This value is divided by 120 and assigned to the `delta` variable, which is then passed onto the `scrollBall()` function.

The `scrollBall()` function manipulates the vertical position of the ball:

```
var currentTop = 1;
var INC = 8
function scrollBall(delta) {
        currentTop = document.getElementById('blueDot').offsetTop;
        if (delta < 0)
                    currentTop = currentTop - INC;
        else if (delta > 0)
                currentTop = currentTop + INC;
        if (currentTop > 390)
                currentTop = 390;
        else if (currentTop < 1 )
                currentTop = 1;
    document.getElementById('blueDot').style.top = currentTop + 'px';
    setTimeout(function() {
                    window.scrollTo(0, 1);
        }, 100);
}
```

The `currentTop` variable stores the current `top` position of the `blueDot img`. The `delta` variable is then evaluated. If the number is less than 0, `currentTop` decreases by the value of `INC`. If it's greater than 0, it increases by the same amount. Although `INC` can be any value, 8 seems to be the most natural for touch interaction in this example. To ensure the `blueDot` does not scroll off the top or bottom of the viewport, the `currentTop` value is evaluated and adjusted as needed. The `blueDot style.top` property is updated to the new value. Finally, to ensure that inadvertent touch inputs do not cause the URL bar to display, `window.scrollTo()` is called.

This technique enables you to effectively utilize the two-finger scroll in your own applications. However, using this touch input has two caveats:

❑   The biggest downfall to implementing the two-finger scroll in your application is that it is a tricky touch input for a user to pull off consistently. If one of the fingers lifts up off the glass surface, Safari is unforgiving. It immediately thinks the user is performing a one-finger scroll and begins to scroll the entire page.

❑   It's impossible to effectively program a flicking action in association with a two-finger scroll to accelerate the rate of movement of the element you are manipulating. Instead, there is always a 1:1 correspondence between the firing of a `mousescroll` event and the position of the element.

If you need to trap for scrolling actions that go beyond the basic two-finger scrolls, use touch or gesture events instead.

Finally, I should mention that this demo works only in portrait mode and is not enabled for landscape.

Touch Events

In addition to the mouse-emulation events, Safari on iPhone captures an additional touch-related sequence of events for each finger that touches the screen surface. A touch event begins when a finger touches the screen surface, continues when the finger moves, and ends when the finger leaves it. The four touch events are shown in Table 7-5.

Table 7-5: Touch Events

Event	Description
`touchstart`	Fires when a finger touches the screen surface
`touchmove`	Fires when the same finger moves across the surface
`touchend`	Fires when a finger leaves the surface of the screen
`touchcancel`	Fires when the OS cancels the touch

Each of these touch events has properties that enable you to get information about the touch:

- ❑ `event.touches` returns an array of all the touches on a page. For example, if the surface had four fingers, the `touches` property would return four `Touch` objects.
- ❑ `event.targetTouches` returns just the touches that were started from the same element on a page.
- ❑ `event.changedTouches` returns all the touches involved in the event.

The information contained in each of these properties varies depending on what multitouch event is performed. Table 7-6 shows several scenarios.

Table 7-6: Touch Scenarios

Touch Scenario	Touches	Target Touches	Changed Touches
One finger touches surface	1 item	1 item	1 item
Two fingers on the same target at the same time	2 items	2 items	2 items
Two fingers touching different targets at different times	2 items	1 item	1 item
Moving one finger across the surface			1 item
Moving two fingers across the surface			2 items
Lift up a finger from the surface	(Removed from list)	(Removed from list)	1 item

The items inside these arrays are `Touch` objects, which provide the information related to the touch event shown in Table 7-7.

Table 7-7: Touch Object Properties

Property	Description
clientX	x coordinate of the object relative to the full viewport (not including scroll offset)
clientY	y coordinate relative to the full viewport (not including scroll offset)
identifier	Unique integer of the touch event
pageX	x coordinate relative to the page
pageY	y coordinate relative to the page
screenX	x coordinate relative to the screen
screenY	y coordinate relative to the screen
target	Originating element (node) that dispatched the touch event

Unlike mouse-emulated events, you can trap for multiple touches to occur on-screen at the same time and then have your app respond to them.

For example, if you want to do a test to determine how these events are fired, you can add event handlers and then take different actions when these events occur. Here's the JavaScript code:

```
function init()
{
    document.addEventListener("touchstart", touchEventHandler, false);
    document.addEventListener("touchmove", touchEventHandler, false);
    document.addEventListener("touchcancel", touchEventHandler, false);
    document.addEventListener("touchend", touchEventHandler, false);
}


function touchEventHandler(event)
{
    // Gather basic touch info
    var numTouch = event.touches.length;
    var numTargetTouches = event.targetTouches.length;
    var numChangedTouches = event.changedTouches.length;

    // Get first touch object
    if (numTouch > 0)
    {
```

```
        var touchObj = event.touches[0];
        var x = touchObj.screenX;
        var y = touchObj.screenY;
    }

    if (event.type == "touchstart")
    {
        // do something to begin
    }
    else if (event.type == "touchmove")
    {
        // do something on move
    }
    else if (event.type == "touchend")
    {
        // do something to end
    }
    else
    {
        // do something when cancelled
    }
}
```

Gesture Events

During a multitouch sequence, touch events are dispatched through `touchstart`, `touchmove`, `touchend`, and `touchcancel`. However, Safari also dispatches gesture events when multiple fingers are touching the surface of the screen. The three gesture events are shown in Table 7-8.

Table 7-8: Touch Events

Event	Description
Gesturestart	Fires when two or more fingers touch the screen surface
Gesturechange	Fires when these fingers move across the surface or perform another change
Gestureend	Fires when one of the fingers involved in the gesture leaves the surface of the screen

The two key properties associated with gesture events are as follows:

❑ `scale` returns the multiplier of the pinch or push since the gesture started. (`1.0` is the baseline value.)

❑ `rotation` returns the rotation value since the gesture began.

Orientation Change

Users of your application are free to rotate their iPhone or iPod touch in their hands at anytime. Therefore, one of the most important events that you need to account for in your application is properly responding to these events no matter where they are in your application. In this section, I'll show you how to detect an orientation change, how to style for both portrait and landscape modes, how to tweak the UI based on the current context, and how to work with older versions of Safari on iPhone.

Detecting an Orientation Change

One of the unique events that an iPhone Web application developer needs to be able to trap for is the change between vertical and horizontal orientation. Safari (iPhone OS 1.1.1 and later) provides support for the `orientationchange` event handler of the `window` object. This event is triggered each time the user rotates the device. The following code shows how to configure the `orientationchange` event:

```
<!DOCTYPE html PUBLIC "-//W3C//DTD XHTML 1.0 Strict//EN"
          "http://www.w3.org/TR/xhtml1/DTD/xhtml1-strict.dtd">
<html xmlns="http://www.w3.org/1999/xhtml">
<head>
<title>Orientation Change Example</title>
<meta name="viewport" content="width=320; initial-scale=1.0; maximum-scale=1.0;
user-scalable=0;">
<script language="javascript" type="text/javascript">
    function orientationChangeHandler()
    {
      var str = "Orientation: ";
      switch(window.orientation)
      {
          case 0:
              str += "Portrait";
          break;

          case -90:
              str += "Landscape (right, screen turned clockwise)";
          break;

          case 90:
              str += "Landscape (left, screen turned counterclockwise)";
          break;

          case 180:
            str += "Portrait (upside-down portrait)";
          break;
      }
      document.getElementById("mode").innerHTML = str;
    }
</script>
</head>
<body onload="orientationChangeHandler();" onorientationchange=
"orientationChangeHandler();">
```

```
<h4 id="mode">Ras sed nibh.</h4>
<p>
Donec semper lorem ac dolor ornare interdum. Praesent condimentum. Suspendisse
lacinia interdum augue. Nunc venenatis ipsum sed ligula. Aenean vitae lacus. Sed
sit amet neque. Vestibulum ante ipsum primis in faucibus orci luctus et ultrices
posuere cubilia Curae; Duis laoreet lorem quis nulla. Curabitur enim erat,
gravida ac, posuere sed, nonummy in, tortor. Donec id orci id lectus convallis
egestas. Duis ut dui. Aliquam dignissim dictum metus.
</p>
</body>
</html>
```

An `orientationchange` attribute is added to the `body` element and assigned the JavaScript function `orientationChangeHandler()`. The `orientationChangehandler()` function evaluates the `window.orientation` property to determine the current state: `0` (Portrait), `-90` (Landscape, clockwise), `90` (Landscape counterclockwise), or `180` (Portrait, upside down). The current state string is then output to the document.

However, note that the `orientationchange` event is not triggered when the document loads. Therefore, to evaluate the document orientation, assign the `orientationChangeHandler()` function to the `load` event.

Changing a Style Sheet When the Orientation Changes

The most common procedure that iPhone developers will want to use an `orientationChange` handler for is to specify a style sheet based on the current viewport orientation. To do so, you can expand upon the previous `orientationChangeHandler()` function by updating the `orient` attribute of the `body` element based on the current orientation and then updating the active CSS styles off that attribute value.

To add this functionality, you begin with a basic XHTML document. The following code, based on a liquid layout template by Joe Hewitt (the original developer of iUI), uses a series of `div` elements to imitate a basic iPhone interface, consisting of a top toolbar, a content area, and a bottom toolbar. The content inside the `center` `div` is going to be used for testing purposes only. Here's the code:

```
<!DOCTYPE html PUBLIC "-//W3C//DTD XHTML 1.0 Strict//EN"
        "http://www.w3.org/TR/xhtml1/DTD/xhtml1-strict.dtd">
<html xmlns="http://www.w3.org/1999/xhtml">
<head>
<title>Change Stylesheet based on Orientation</title>
<meta name="viewport" content="width=320; initial-scale=1.0; maximum-scale=1.0;
user-scalable=0;">
</head>
<body>
    <div id="canvasMain" class="container">
        <div class="toolbar anchorTop">
            <div class="main">
                <div class="header">AppTop</div>
            </div>
        </div>
        <div class="center">
```

```
            <p>Orientation mode:<span id="iMode"></span></p>
            <p>Width:<span id="iWidth"></span></p>
            <p>Height:<span id="iHeight"></span></p>
            <p>Bottom toolbar height:<span id="iToolbarHeight"></span></p>
            <p>Bottom toolbar top:<span id="iToolbarTop"></span></p>
            </div>
            <div id="bottomToolbar" class="toolbar anchorBottom">
                <div class="main">
                    <div class="header">
                        AppBottom
                    </div>
                </div>
            </div>
        </div></body>
    </html>
```

Next, add CSS rules to the document head. However, notice that the selector for the final four rules (highlighted below) depends on the state of the `orient` attribute of body:

```css
<style type="text/css" media="screen">
    body {
        margin: 0;
        padding: 0;
        width: 320px;
        height: 416px;
        font-family: Helvetica;
        -webkit-user-select: none;
        cursor: default;
        -webkit-text-size-adjust: none;
    background: #000000;
    color: #FFFFFF;
    }
    .container {
            position: absolute;
            width: 100%;
    }
    .toolbar {
            position: absolute;
            width: 100%;
            height: 60px;
            font-size: 28pt;
    }
    .anchorTop {
            top: 0;
    }
    .anchorBottom {
            bottom: 0;
    }
    .center {
            position: absolute;
            top: 60px;
            bottom: 60px;
```

```css
    }
    .main {
        overflow: hidden;
        position: relative;
    }
    .header {
        position: relative;
        height: 44px;
        -webkit-box-sizing: border-box;
        box-sizing: border-box;
        background-color: rgb(111, 135, 168);
        border-top: 1px solid rgb(179, 186, 201);
        border-bottom: 1px solid rgb(73, 95, 144);
        color: white;
        font-size: 20px;
        text-shadow: rgba(0, 0, 0, 0.6) 0 -1px 0;
        font-weight: bold;
        text-align: center;
        line-height: 42px;
    }
/* Styles adjusted based on orientation  */
    body[orient='portrait'].container {
        height: 436px;
    }
    body[orient='landscape'].container {
        height: 258px;
    }
    body[orient='landscape'].toolbar {
        height: 30px;
        font-size: 16pt;
    }
    body[orient='landscape'].center {
        top: 50px;
        bottom: 30px;
    }
</style>
```

Based on the `body` element's `orient` value, the `container` CSS class changes its height, the top and bottom toolbars adjust their `height` and `font-size`, and the main content area (the `center` class) is repositioned to fit with the sizing changes around it.

With the XHTML and CSS styles in place, you are ready to add the `orientationchange` handler to the `body` element:

```html
<body onorientationchange="orientationChangeHandler()"
onload="orientationChangeHandler()">
```

Next, you can add JavaScript code inside the document head:

```html
<script language="javascript" type="text/javascript">
```

```javascript
function orientationChangeHandler()
{
    if (window.orientation == 0 || window.orientation == 180)
        document.body.setAttribute('orient', 'portrait')
    else
        document.body.setAttribute('orient', 'landscape');
    document.getElementById('iMode').innerHTML =
        document.body.getAttribute('orient');
    // currentWidth is a global variable defined in the document head
    document.getElementById('iWidth').innerHTML = currentWidth + 'px';
    document.getElementById('iHeight').innerHTML =
        document.getElementById('canvasMain').offsetHeight + 'px';
    document.getElementById('iToolbarHeight').innerHTML =
        document.getElementById('bottomToolbar').offsetHeight +'px';
    document.getElementById('iToolbarTop').innerHTML =
        document.getElementById('bottomToolbar').offsetTop +'px';
}
</script>
```

The `orientationChangeHandler()` function is called when the window loads or changes orientation. It updates the `body` element's `orient` attribute to either `portrait` or `landscape`.

This example also outputs some of the changing `div` size and position values into a series of `span` elements for information purposes.

Figures 7-2 and 7-3 show the document loaded in both portrait and landscape modes, respectively.

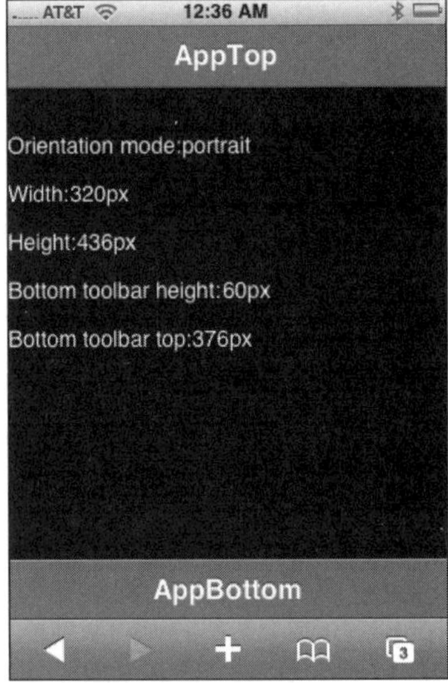

Figure 7-2: Portrait mode

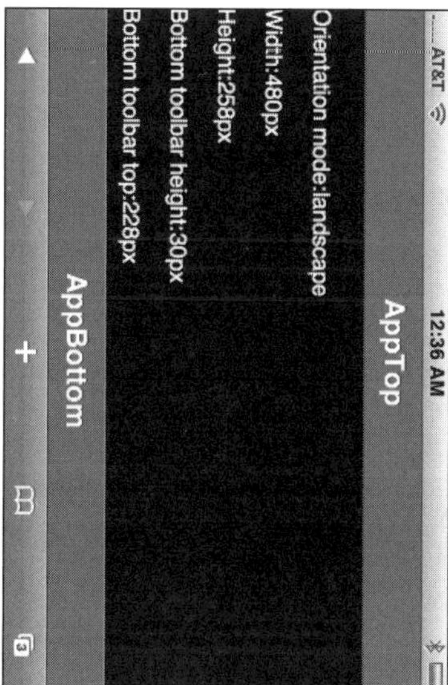

Figure 7-3: Landscape mode

Changing Element Positioning Based on Orientation Change

Once you understand the basic interaction between an `orientationChangeHandler()` function and orientation-dependent styles, you can begin to dynamically position elements of the UI based on whether the current viewport is in portrait or landscape mode. Suppose, for example, you want to align an arrow image to the bottom-left side of a page. Here's the `img` declaration:

```
<img id="pushBtn" src="bottombarknobgray.png"/>
```

To align the graphic in portrait mode, you can specify the CSS rule as follows:

```
#pushbtn
{
    position: absolute;
    left: 10px;
    top: 360px;
}
```

However, if you leave the positioning as is, the button goes off screen when the user tilts the viewport to landscape mode. Therefore, a second landscape-specific rule is needed for the button image, with an adjusted `top` value:

```
body[orient="landscape"] #pushBtn
{
    left: 10px;
    top: 212px;
}
```

The `orientationChangeHandler()` function is as follows:

```
function orientationChangeHandler()
{
    if (window.orientation == 0 || window.orientation == 180)
        document.body.setAttribute('orient', 'portrait')
    else
        document.body.setAttribute('orient', 'landscape');
}
```

As Figures 7-4 and 7-5 show, the button image aligns to the bottom left of the page document in both portrait and landscape modes, respectively.

Figure 7-4: Push button aligned in portrait mode

Figure 7-5: Push button aligned in landscape mode

Trapping for Key Events with the On-Screen Keyboard

As with an ordinary Web page, you can validate keyboard input by trapping the keydown event. To illustrate, suppose you have an input field in which you want to prevent the user from entering a numeric value. To trap for this, begin by adding a keydown handler to the input element:

```
<input onkeydown="return validate(event)" />
```

In the document header, add a `script` element with the following code inside:

```
function validate(e) {
    var keynum = e.which;
    var keychar = String.fromCharCode(keynum);
    var chk = /\d/;
    return !chk.test(keychar)
}
```

As a standard JavaScript validation routine, this function tests the current character code value to determine whether it is a number. If a non-number is found, `true` is returned to the `input` field. Otherwise, `false` is sent back and the character is disallowed.

Summary

Given iPhone's touch interface, event handling is a key part of any iPhone application, whether it is native or Web-based. This chapter focused on event handling in iPhone Web apps. I began by exploring the three major types of touch events. The first type of event occurs when a single- or double-finger action emulates a mouse. The second type is a touch event, which begins when a finger touches the screen surface, continues when the finger moves, and ends when the finger leaves it. The third type is a gesture event, which combines a set of touch events into a "gesture". The chapter than concentrated on an event that is unique to a mobile device such as iPhone and iPod touch: the change in orientation when a user rotates the device in their hands.

8

Programming the Canvas

The unique platform capabilities of the iPhone enable developers to create innovative applications inside of Safari that go beyond the normal "Web app" fare. Safari/WebKit's support for the `canvas` element opens drawing and animation capabilities in an ordinary HTML page that was previously available only by using Flash or Java. The `canvas` element is part of the Web Hypertext Application Technology Working Group (WHATWG) specification for HTML 5.0 (or HTML5).

However, once you begin to open these capabilities, you need to be sure that you are working with an iPhone and iPod touch rather than a standard desktop browser that may or may not provide HTML5 support. So, I'll start by showing you how to identify the user agent for iPhone and iPod touch.

Identifying the User Agent

When you are trying to identify the capabilities of the browser requesting your Web site or application, you generally want to avoid detecting the user agent and use object detection instead. However, if you are developing an application designed exclusively for iPhone or need to guarantee the browser being used, user agent detection is a valid option. Therefore, this chapter assumes you are creating a Safari-specific application.

The Safari user agent string for iPhone closely resembles the user agent for Safari on other platforms. However, it contains an iPhone platform name and the mobile version number. Depending on the version of Mobile Safari, it will look something like this:

```
Mozilla/5.0 (iPhone; U; CPU like Mac OS X; en) AppleWebKit/420+ (KHTML, like
Gecko) Version/3.0 Mobile/1A543a Safari/419.3
```

Here's a breakdown of the various components of the user agent:

❑ **The platform string: `(iPhone; U; CPU like Mac OS X; en)`.** Notice the "`like Mac OS X`" line, which reveals some of the underpinnings of the iPhone.

- ❑ **The WebKit engine build number: `AppleWebKit/420+`.** This Safari version number is provided on all platforms (including Mac and Windows).

- ❑ **The marketing version: (`Version/3.0`).** This Safari version number is provided on all platforms (including Mac and Windows).

- ❑ **OS X build number: `Mobile/1A543a`.**

- ❑ **Safari build number: `Safari/419.3`.**

The iPod touch user agent is similar, but it is distinct from `iPod` as the platform:

```
Mozila/5.0 (iPod; U; CPU like Mac OS X; en) AppleWebKit/420.1 (KHTML, like
Gecko) Version/3.0 Mobile/3A101a Safari/419.3
```

The version numbers will change, obviously, when Apple updates Safari, but the string structure will stay the same.

To test whether the device is an iPhone/iPod touch, you need to perform a string search on `iPhone` and `iPod`. The following function returns `true` if the user agent is either an iPhone or iPod touch:

```
function isAppleMobile() {
   result ((navigator.platform.indexOf("iPhone") != -1) ||
          (navigator.userAgent.indexOf("iPod") != -1))
}
```

Be sure not to test for the string `Mobile` within the user agent, because a non-Apple mobile device (such as Nokia) might be based on the WebKit-based browser.

If you need to do anything beyond basic user agent detection and test for specific devices or browser versions, however, consider using WebKit's own user agent detection script available for download at `trac.webkit.org/projects/webkit/wiki/DetectingWebKit`. By linking `WebKitDetect.js` to your page, you can test for specific devices (iPhone and iPod touch) as well as software versions. Here's a sample detection script:

```
<!DOCTYPE html PUBLIC "-//W3C//DTD XHTML 1.0 Strict//EN"
          "http://www.w3.org/TR/xhtml1/DTD/xhtml1-strict.dtd">
<html xmlns="http://www.w3.org/1999/xhtml">
<head>
<title>User Agent Detection via WebKit Script</title>
<meta name="viewport" content="width=320; initial-scale=1.0;
maximum-scale=1.0; user-scalable=0;">
<script type="application/x-javascript" src="WebKitDetect.js"></script>
</head>
<body>
<p id="log"></p>
</body>
<script type="application/x-javascript">
function addTextNode(str) {
  var t = document.createTextNode(str);
```

```
      var p = document.getElementById("log");
      p.appendChild(t);
    }
    if ( WebKitDetect.isMobile() ) {
      var device = WebKitDetect.mobileDevice();
      // String found in Settings/General/About/Version
      var minSupport = WebKitDetect.mobileVersionIsAtLeast("1C28");
      switch( device ) {
        case 'iPhone':
          if ( minSupport ) {
            addTextNode('If this were a real app, I would launch its URL right now.');
          }
          else {
            addTextNode('Please upgrade your iPhone to the latest update before
            running this application.');
          }
          break;
        case 'iPod':
          addTextNode('If this were a real app, I would launch its iPod touch
          version.');
          break;
        default:
          addTextNode( 'This mobile device is not supported by this application.
    Go to your nearest Apple store and get an iPhone.');
          break;
      }
    }
    else {
      addTextNode( 'Desktop computers are so 1990s. Go to your nearest Apple store
      and get an iPhone.' );
    }
</script>
</html>
```

With the `WebKitDetect.js` script included, the `WebKitDetect` object is accessible. Begin by calling its `isMobile()` method to determine whether the device is a mobile one. Next, check to ensure that the mobile version is the latest release, and save that result in the `minSupport` variable. The `switch` statement then evaluates the mobile device. If it is an iPhone, `switch` checks to see if `minSupport` is `true`. If so, a real application would begin here. If `minSupport` is `false`, the user is notified to update his or her iPhone to the latest software version. The remaining two `case` statements evaluate for an iPod touch or an unknown mobile device. The final `else` statement is called if the device is not a mobile computer.

Programming the iPhone Canvas

C++ and other traditional software programmers have long worked with a *canvas* on which to draw graphics. In contrast, Web developers typically have programmed the presentation layer using HTML and CSS. But unless they used Flash or Java, they had no real way to actually draw graphical content on a Web page. However, both desktop and mobile versions of Safari support the `canvas` element to provide a resolution-dependent bitmap region for drawing arbitrary content. The `canvas` element defines a drawing region on your Web page that you then draw on using a corresponding JavaScript `canvas` object.

The canvas frees you up as an application developer to not only draw anything you want to, but use the canvas as a way to render graphs, program games, or add special effects. On Mac OS X, the canvas is often used for creating Dashboard widgets. On iPhone, Apple uses the canvas for both the Clock and Stocks built-in applications.

Canvas programming can be a mindset difference for Web developers used to manipulating existing graphics rather than creating them from scratch. It is the loose equivalent of a Photoshop expert beginning to create content using an Adobe Illustrator–like program in which all the graphics are created in a nonvisual manner.

Defining the Canvas Element

Think of a canvas as a rectangular block region of a page that you have full control over what gets drawn on it. The canvas is defined using the `canvas` element:

```
<canvas id="theCanvas" width="300" height="300"/>
```

Except for the `src` and `alt` attributes, the `canvas` element supports all the same attributes as the `img` tag. The `id`, `width`, and `height` attributes are not required, but they should be defined as a sound programming practice. The `width` and `height` are usually defined in pixels, although they could also be a percentage of the viewport.

You can place multiple `canvas` elements on a page, just as long as each one has its own unique ID.

Getting a Context

Once a canvas region is defined on your Web page, you can draw inside of the flat two-dimensional surface using JavaScript. Just like a Web page, the canvas has an origin (0,0) in the top-left corner. By default, all the *x,y* coordinates you specify are relative to this position.

As the first step in working with the canvas, you need to get a `2d context` object. This object, which is responsible for managing the canvas's graphics state, is obtained by calling the `getContext()` method of the `canvas` object:

```
var canvas = document.GetElementById("theCanvas");
var context = canvas.getContext("2d");
```

Or, because you don't normally work with the `canvas` object directly, you can also combine the two lines:

```
var context = document.GetElementById("theCanvas").getContext("2d");
```

All the drawing properties and methods you work with are called from the `context` object. The `context` object has many properties (see Table 8-1) that determine how the drawing looks on the page.

Table 8-1: Context Properties

Property	Description
fillStyle	Provides the CSS color or style (gradient, pattern) of the fill of a path.
font	Specifies the font used.
globalAlpha	Specifies the level of transparency of content drawn on the canvas. The floating point value is between 0.0 (fully transparent) and 1.0 (fully opaque).
globalCompositeOperation	Specifies the compositing mode to determine how the canvas is displayed relative to background content. Values include copy, darker, destination-atop, destination-in, destination-out, destination-over, lighten, source-atop, source-in, source-out, source-over, and xor.
lineCap	Defines the end style of a line. String values include butt for flat edge, round for rounded edge, square for square ends. (Defaults to butt.)
lineJoin	Specifies the way lines are joined. String values include round, bevel, and miter. (Defaults to miter.)
lineWidth	Specifies the line width. The floating point value is greater than 0.
miterLimit	Specifies the miter limit for drawing a juncture between line segments.
shadowBlur	Defines the width that a shadow covers.
shadowColor	Provides CSS color for the shadow.
shadowOffsetX	Specifies the horizontal distance of the shadow from the source.
shadowOffsetY	Specifies the vertical distance of the shadow from the source.
strokeStyle	Defines the CSS color or style (gradient, pattern) when stroking paths.
textAlign	Determines the text alignment.
textBaseline	Specifies the baseline of the text (top, hanging, middle, alphabetic, ideographic, or bottom)

Drawing a Simple Rectangle

There are several techniques for drawing on the canvas. Perhaps the most straightforward is by drawing a rectangle. To do so, you work with three context methods:

- ❑ `context.fillRect(x,y,w,h)` draws a filled rectangle.
- ❑ `context.strokeRect(x,y,w,h)` draws a rectangular outline.
- ❑ `context.clearRect(x,y,w,h)` clears the specified rectangle and makes it transparent.

For example, suppose you would like to draw a rectangular box with a set of squares inside of it and a rectangular outline on the outside. Here's a JavaScript function that draws that shape:

```
function draw()
{
    var context = document.getElementById('myCanvas').getContext('2d');
    context.strokeRect(10,10,150,140);
    context.fillRect(15,15,140,130);
    context.clearRect(30,30,30,30);
    context.clearRect(70,30,30,30);
    context.clearRect(110,30,30,30);
    context.clearRect(30,100,30,30);
    context.clearRect(70,100,30,30);
    context.clearRect(110,100,30,30);
}
```

Once the `context` is obtained, `strokeRect()` creates a rectangular outline starting at the coordinate (10,10) and is 150 × 140 pixels in size. The `fillRect()` method paints a 140 × 130 rectangle starting at coordinate (15,15). The six `clearRect()` calls clear areas previously painted by `fillRect()`. Figure 8-1 shows the result.

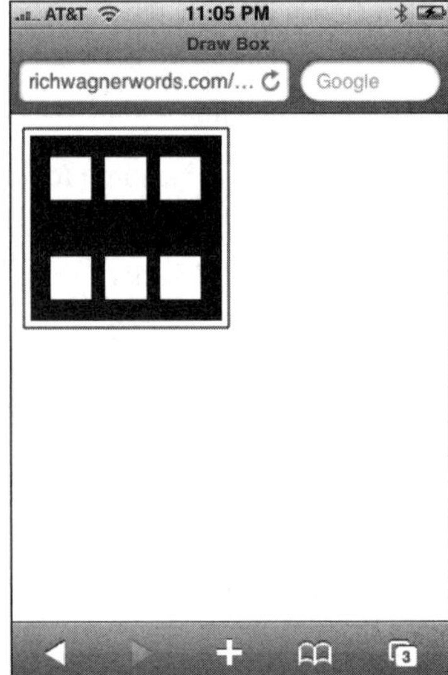

Figure 8-1: Rectangular blocks drawn on a canvas

The full page source is shown in the following code:

```
<!DOCTYPE html PUBLIC "-//W3C//DTD XHTML 1.0 Strict//EN"
         "http://www.w3.org/TR/xhtml1/DTD/xhtml1-strict.dtd">
<html xmlns="http://www.w3.org/1999/xhtml">
<head>
<title>Draw Box</title>
<meta name="viewport" content="width=320; initial-scale=1.0; maximum-scale=1.0;
```

```
user-scalable=0;">
<script type="application/x-javascript">
function draw()
{
    var context = document.getElementById('myCanvas').getContext('2d');
    context.strokeRect(10,10,150,140);
    context.fillRect(15,15,140,130);
    context.clearRect(30,30,30,30);
    context.clearRect(70,30,30,30);
    context.clearRect(110,30,30,30);
    context.clearRect(30,100,30,30);
    context.clearRect(70,100,30,30);
    context.clearRect(110,100,30,30);
}
</script>
</head>
<body onload="draw()">
<canvas id="myCanvas" width="300" height="300" style="position:absolute;
left:0px; top:0px; z-index:1"/>
</body>
</html>
```

Drawing Other Shapes

Nonrectangular shapes are drawn by creating a path for that shape and then either *stroking* (drawing) a line along the specified path or *filling* (painting) in the area inside the path. Much like an Etch A Sketch drawing, paths are composed of a series of *subpaths,* such as a straight line or an arc that together form a shape.

When you work with paths, the following methods are used for drawing basic shapes:

❏   `beginPath()` creates a new path in the canvas and sets the starting point to the coordinate (0,0).

❏   `closePath()` closes an open path and attempts to draw a straight line from the current point to the starting point of the path. The use of `closePath()` is optional.

❏   `stroke()` draws a line along the current path.

❏   `fill()` closes the current path and paints the area within it. (Because `fill()` closes the path automatically, you don't need to call `closePath()` when you use it.)

❏   `lineTo(x,y)` adds a line segment from the current point to the specified coordinate.

❏   `moveTo(x,y)` moves the starting point to a new coordinate specified by the *x,y* values.

Using these methods, you can create a list of subpaths to form a shape. For example, the following code creates two triangles next to each other; one is empty and one is filled. An outer rectangle surrounds both triangles. Here's the code:

```
function drawTriangles()
{
    var context = document.getElementById('myCanvas').getContext('2d');
    // Empty triangle
    context.beginPath();
    context.moveTo(10,10);
```

```
context.lineTo(10,75);
context.lineTo(100,40);
context.lineTo(10,10);
context.stroke();
context.closePath();

// Filled triangle
context.beginPath();
context.moveTo(110,10);
context.lineTo(110,75);
context.lineTo(200,40);
context.lineTo(110,10);
context.fill();
context.closePath();

// Outer rectangle
context.strokeRect(3,3,205,80);
}
```

Figure 8-2 shows the results.

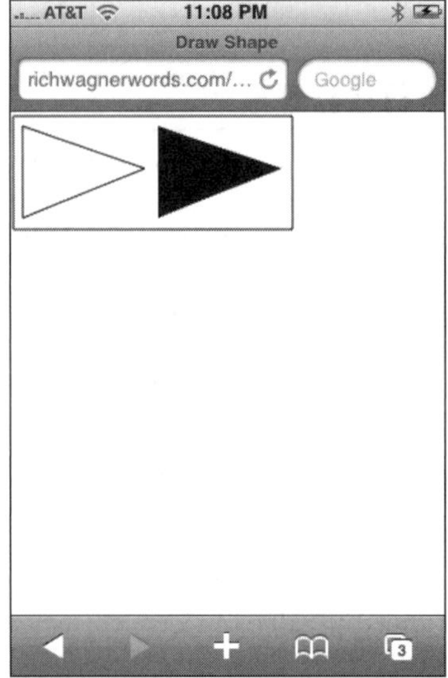

Figure 8-2: Drawing two triangles

If you are new to canvas programming, drawing complex shapes on the canvas can take some getting used to. You may find it helpful initially to go low tech and use a piece of graph paper to sketch out the shapes you are trying to draw and calculate the x,y coordinates using the paper grid.

The JavaScript canvas enables you to go well beyond drawing with straight lines, however. You can use the following methods to create more advanced curves and shapes:

❑   `arc(x, y, radius, startAngle, endAngle, clockwise)` adds an arc to the current sub-path using a radius and specified angles (measured in radians).

- ❏ `arcTo(x1, y1, x2, y2, radius)` adds an arc of a circle to the current subpath by using a radius and tangent points.

- ❏ `quadratricCurveTo(cpx, cpy, x, y)` adds a quadratic Bezier curve to the current subpath. It has a single control point (the point outside of the circle that the line curves toward) represented by cpx, cpy. The *x,y* values represent the new ending point.

- ❏ `bezierCurveTo(cp1x, cp1y, cp2x, cp2y, x, y)` adds a cubic Bezier curve to the current subpath using two control points.

Using `arc()`, I can create a filled circle inside of an empty circle using the following code:

```
function drawCircles()
{
    var context = document.getElementById('myCanvas').getContext('2d');

    // Create filled circle
    context.beginPath();
    context.arc(125,65,30,0, 2*pi, 0);
    context.fill();

    // Create empty circle
    context.beginPath();
    context.arc(125,65,35,0, 2*pi, 0);
    context.stroke();
    context.closePath();
}
```

The `arc()` method starts the arc shape at coordinate (125,65) and draws a 30px radius starting at 0 degrees and ending at 360 degrees at a counterclockwise path.

Figure 8-3 displays the circle shapes that are created when this script is run.

Figure 8-3: Using arc() to draw a circle

Drawing an Image

In addition to lines and other shapes, you can draw an image onto your canvas by using the `drawIm-age()` method. The image can reference either an external image or another `canvas` element on the page. There are actually three ways in which you can call this method. The first variant simply draws an image at the specified coordinates using the size of the image:

```
context.drawImage(image, x, y)
```

The second method enables you to specify the dimensions of the image with the `width` and `height` arguments:

```
context.drawImage(image, x, y, width, height)
```

To do a basic image draw, define the `Image` object and assign an `src`. Next, you want to draw the image, but only after you are certain the image is fully loaded. Therefore, you place the `drawImage()` method inside the image's `onload` handler:

```
<!DOCTYPE html PUBLIC "-//W3C//DTD XHTML 1.0 Strict//EN"
        "http://www.w3.org/TR/xhtml1/DTD/xhtml1-strict.dtd">
<html xmlns="http://www.w3.org/1999/xhtml">
<head>
<title>Draw Image</title>
<meta name="viewport" content="width=320; initial-scale=1.0; maximum-scale=1.0;
user-scalable=0;">
<script type="application/x-javascript">
function drawImg()
{
    var context = document.getElementById('myCanvas').getContext('2d');
    var img = new Image();
    img.src = 'images/beach.jpg';
    img.onload = function() {
      context.drawImage( img, 0, 0 );
    }
}
</script>
</head>
<body onload="drawImg()">
<canvas id="myCanvas" width="300" height="300" style="position:absolute;
left:0px; top:0px; z-index:1"/>
</body>
</html>
```

Figure 8-4 shows the image displayed inside the canvas. Keep in mind that this is not an HTML `img` element, but the external image file drawn onto the context of the canvas.

Additionally, there is a final `drawImage()` option that is slightly more complex:

```
context.drawImage(image, sourcex, sourcey, sourceWidth, sourceHeight, destx,
desty, destWidth, destHeight)
```

Figure 8-4: Drawing an image
onto the canvas

In this variant, the method draws a subsection of the image specified by the source rectangle
(`sourcex`, `sourcey`, `sourceWidth`, `sourceHeight`) onto a destination rectangle specified by the
final arguments (`destx`, `desty`, `destWidth`, and `destHeight`). For example, suppose you just
wanted to display the rock thrower in Figure 8-4 rather than the entire picture. Using this expanded
syntax of `drawImage()`, you want to extract a 79 × 131px rectangle from the original picture starting
at the coordinate (151,63). You then paint the same sized rectangle at coordinate (10,10) on the canvas.
Here is the updated code:

```
function drawImg(){
   var canvas = document.getElementById('myCanvas');
   var context = canvas.getContext('2d');
   var img = new Image();
   img.src = 'images/beach.jpg';
   img.onload = function() {
      context.drawImage( img, 151, 63, 79, 131, 10, 10, 79, 131 );
   }
}
```

Figure 8-5 shows the result.

Suppose you had occasion to create a self-contained web app with no dependencies on external files.
If so, you could also use a `data:` URI encoded image to completely eliminate the need for an external
image file for canvas painting. For example, start with an online image encoder, such as the one available at `www.scalora.org/projects/uriencoder`. Using this tool, you encode the image, as shown
in Figure 8-6.

Figure 8-5: Painting a portion of an image

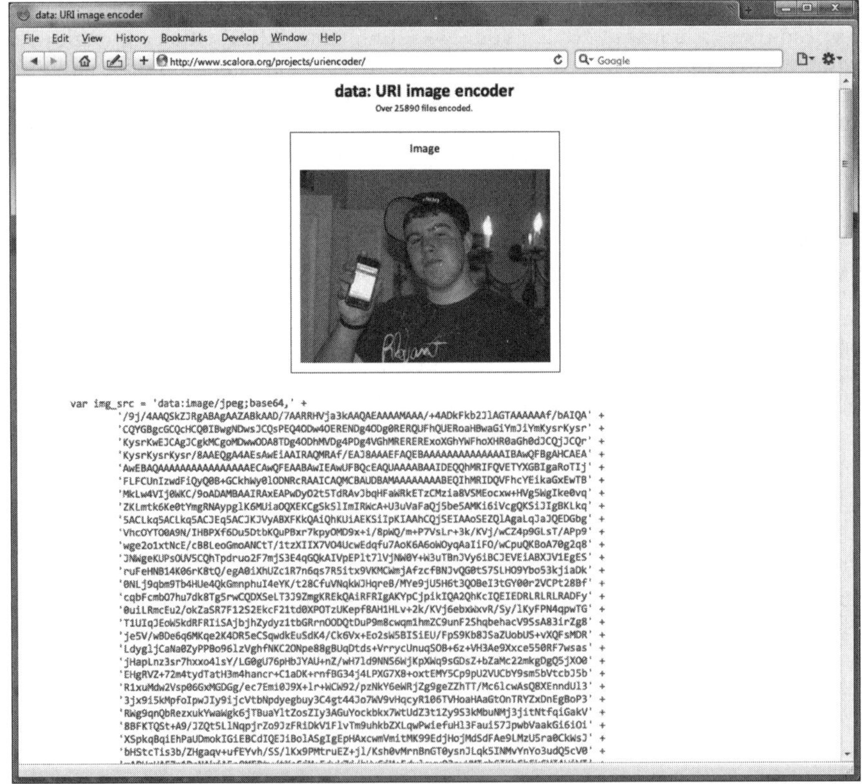

Figure 8-6: Encoding an image

You can then integrate the outputted encoded string into the script code as the `Image` object's source. (Much of the encoded text for this example has been removed for space reasons.)

```
function drawImg(){
var img_src = 'data:image/jpeg;base64,' +
'/9j/4AAQSkZJRgABAgAAZABkAAD/7AARRHVja3kAAQAEAAAAMAAA/+4ADkFkb2JlAGTAAAAAf/bAIQA'+
'CQYGBgcGCQcHCQ0IBwgNDwsJCQsPEQ4ODw4OERENDg4ODg0RERQUFhQUERoHBwaGiYmJiYmKysrKysr'+
'KysrKwEJCAgJCgkMCgoMDwwODA8TDg4ODhMVDg4PDg4VGhMRERERExoXGhYWFhoXHR0aGh0dJCQjJCQr'+
'KysrKysrKysr/8AAEQgA4AEsAwEiAAIRAQMRAf/EAJ8AAAEFAQEBAAAAAAAAAAAAIBAwQFBgAHCAEA'+
'AwEBAQAAAAAAAAAAAAAAAAECAwQFEAABAwIEAwUFBQcEAQUAAAABAAIDEQQhMRIFQVETYXGBIgaRoTIj'+
'FLFCUnIzwdFiQyQ0B+GCkhWy0lODNRcRAAICAQMCBAUDBAMAMAAAAAAABEQIhMRIDQVFhcYEikaGxEwTB'+
'MkLw4VIj0WKC/9oADAMBAAIRAxEAPwDyO2t5TdRAvJbqHFaWRkETzCMzia8VSMEocxw+HVg5WgIke0vq'+
'ZKLmtk6Ke0tYmgRNAypglK6MUiaOQXEKCgSkSlImIRWcA+U3uVaFaQj5be5AMKi6iVcgQKSiJIgBKLkq'+
'5ACLkq5ACLkq5ACJEq5ACJKJVyABXFKkQAiQhKUiAEKSiIpkIAAhCQjSEIoSEZQlAgaLqJaJQEDGbg'+
'VhcOYTO0A9N/IHBPXf6Du5DtbKQuPBxr7kpyOMD9x+i/8pWQ/m+P7VsLr+3k/KVj/wCZ4p9GLsT/APp9'+
'OmIIzR8p0jsHErPNY1jQxowbgpW5TOudxkfm2LyN7zi5R8u5IlvIJAPYhPvRHNDkfsTEXnpO1tXXcu4X'+
'rOpY7TC67ljOT3g6YYj+Z5CsLL1B6lvPUcTTdyPmmma18IPyg2vnb08tLW1Xendsu7v08+K2idI7cdwg'+
'hl0jAQwMMz9R4CrhmrqT07centnvNyaz6neJ2O1Ob8NvHJ+oWfiIriV2cfHfZV1lJe+zX0+BwcvLx77q'+
'0Ws/9dE/r4Z1M5u8zX7zaQwnRbwuj6IpgOrKZtQH+9Qt7AG87g0ClLmYAf8AyOVptdp1vU9iJcI7SOG4'+
'm1cGQRN1d/4qpguDcbg641o8zPkmdq/FR0gPtWPIpTnG6+PRf3N+PDUZ28cv/wBP+wF8a3cnGhDa9rQG'+
'n7FHfknnwyCBk7h5JXODXE4ktpqw8V13Z3VqY23MToTK3qMDsKtOFfaFg02248fidCaSSnw+AwE5FHJK'+
'8MYKuOPIADMk8AOKft7a7EzIxZOnlkAfHG5j6uacjpbQlpVlJuh28fTRWVtb3rD/AFFB6OsUFCI/nl3eV'+
'VeNRNnCXgTbkcxVbm/H5kaBn09jNcwQNuomPZFLdPJDWudiGRsqCa0zPsUKW7nlFCQxnFjAGtFewLZv9'+
'Y3EOxWt4yxtRNPM+Jw0DSTG0fM0tpj5lRyerdzeySLo2jYpfijFtHT3hbcleNKqXI9JhV7mPFfls7N8S'+
'w4127YfQoySSTmTxS0XONXF1AKmtAKDwCTCq5TqDM0jm6SQRliATQDTnRABRKEqG2wwhF3euK5AAkJT7'+
'lyRACJOK4lCD5jjkgQYxNEkhp5R4pNWkVGaCpOJzKcBIuCRKkOaBirqmlOGdF1SuwSGenAp+U/IPco4K'+
'duHhls9xya0krZHP0Kbc7k9Nlow4yCslPw8vFdZQgDLioUTnTSumdm817hwVrbUHgsnlnVVbak6BoACl'+
'tywUWM44J8O9ipCZz60VLvc/Tt3mvBWssoaM1nN3c65kZA3OV7WD/caIYkZ6/wBpvdvbFJcN8lO3qxyD'+
'I6vNpPaFDIBHby5r1q9261vLF1jcN1RaQ0c2loo1ze0LzDddtn2y8fazCtMWP4PacnBJCsoK84Yc8ihJ'+
'4HNE7t45psk6gDmEyGeibFul7snoqO7tdIDhcyOLhUdR0sUMOHZ5lVbDvO4Xd/fyXtw+461lOJNZqMhS'+
'jchSvBO31yGf4422AHGa5fXuY6R32kLP7XeiyuxM8F0T2vilAz0SN0mnaM12X5HW3Cpaqq1b7ZOKnErU'+
'57bU7O9kn1wzf7fBt9pDPudwDNdbna6YoWsLi2GFjwSup/E6lexQrR8M/pq93G7gbd3AdNBFduYGOdEG'+
'lzHUAwAJIw7lQWPq7dba4MshbdQuPmhkGAYRpc2NwoWVbhhgrrbmnfrXc4LS55kc6b6ZvSdGyJsMQkNWM'+
'axxaaCuQFFfFaU5aXxTWLe3u9ZMr8N+NbrvE19yelU4jwKpkcMVtbXWirLG0bcvPGaSWRsbSPzmvc1X+3 +3'+
'emYtw2fZru9A0NdNeX9w8+YxF3Uax1eD8yqrcd22yz3C92qe1N3tzGQWvy39N4NsDVzTjm5xVZuXqndL'+
'xs9tFK6322UgMtGkUaxrQxrNVK0o3FZ7+Ljb3e6Ft2r/ACX6YNdnNyJbfZL3bn/jZaefuZZ+p9svLG'+
'43 Nhi+rtnTR2keHUERLWMcWA6tJzPYsp61sWTGHeLZ8cwlAivei7W1k1ARjn5gciskCpdhul/t7nus5j'+
'D1BpkAALXDhVrgRUcCov+WuSuy9YT6rv3gvj/D+1ffx2yuj0jtJO3d4isLHbqaZbQPM7eIllDZHA9orp'+
'8FUpQ/W8ulJLnnU5+bqnEnE41SEAU0mtQCcKUPJc17bnPTReSwdVK7axq8tvxewWcVyTilUFnLly5AHJC'+
'lJQlAjkhPtSoCUwEJQB2LkvFu8rimhMPVq8EtUDCKdiJApcqhOaCUwEJQB2LkvGvcrimhMCVq8ktkDSKd'+
'iJjQ4kAJJApxpcphOaCkwEJQB2LkvFvcrSmhMCVq8lOkDSKdiJjQ4kAJJApxpcpkOaCkwEJQB2LkvFu'+
'cCaK7ZRQ8mSNStXgloIkUJbQNBC3qXOcUJwaHGNyBKaAwENpvTpTpY6IFQYJRpRNtQZxKSkODpAyKc/I/9k=';
  var canvas = document.getElementById('myCanvas');
  var context = canvas.getContext('2d');
  var img = new Image();
  img.src = img_src;
  img.onload = function() {
    context.drawImage( img, 10, 10 );
  }
}
```

Figure 8-7 shows the rendered image.

Figure 8-7: Encoded image

Adding Color and Transparency

The `fillStyle` and `strokeStyle` properties of the `context` object provide a way for you to set the color, alpha value, or style of the shape or line you are drawing. (See Table 8-1 for a list of all context properties.) If you would like to set a color value, you can use any CSS color, such as the following:

```
context.fillStyle="#666666";
context.strokeStyle=rgb(125,125,125);
```

Once you set `fillStyle` or `strokeStyle`, it becomes the default value for all new shapes in the canvas until you reassign it.

You can also use `rgba(r,g,b,a)` to assign an alpha value to the shape you are filling in. The r, g, and b parameters take an integer value between 0 and 255, whereas a is a float value between 0.0 and 1.0 (0.0 being fully transparent, and 1.0 being fully opaque). For example, the following code draws two circles in the canvas. The large circle has a 90 percent transparency value, whereas the smaller circle has a 30 percent transparency value:

```
function drawTransCircles()
{
    var context = document.getElementById("myCanvas").getContext("2d");
    // Large circle—90% transparency
    context.fillStyle = "rgba(13,44,50, 0.9)";
    context.beginPath();
    context.arc(95,90,60,0, 2*pi, 0);
    context.fill();
    // Smaller circle—30% transparency
```

```
        context.fillStyle = "rgba(0,0,255, 0.3)";
        context.beginPath();
        context.arc(135,120,40,0, 2*pi, 0);
        context.fill();
    }
```

Figure 8-8 shows the two colored, semitransparent circles. Alternatively, you can declare the `context`
`.globalAlpha` property to set a default transparency value for all stroke or fill styles. Once again, the
value should be a float number between `0.0` and `1.0`.

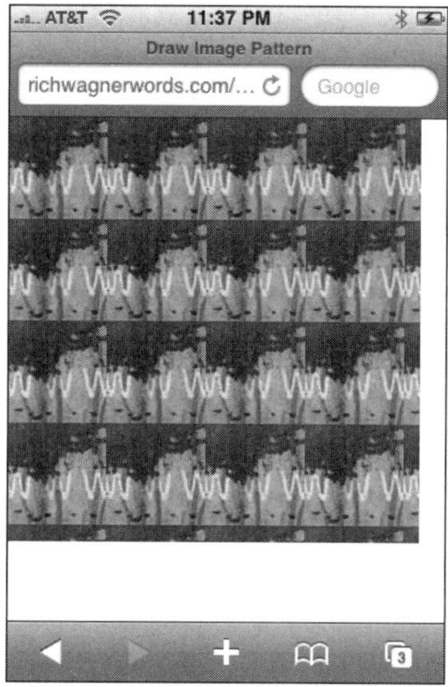

Figure 8-8: Using an image as a fill

Creating an Image Pattern

You can use an external image to create an image pattern on the back of a `canvas` element using the
`createPattern()` method. The syntax is as follows:

```
patternObject = context.createPattern(image, type)
```

The `image` argument references an `Image` object or a different `canvas` element. The `type` argument
is one of the familiar CSS pattern types: `repeat`, `repeat-x`, `repeat-y`, and `no-repeat`. The method
returns a `Pattern` object, as shown in the following example:

```
function drawPattern()
{
    var context = document.getElementById('myCanvas').getContext('2d');
    var pImg = new Image();
    pImg.src = 'images/tech.jpg';
    // call when image is fully loaded
```

```
        pImg.onload = function() {
          var pat = context.createPattern(pImg,'repeat');
          context.fillStyle = pat;
          context.fillRect(0,0,300,300)
        }
    }
```

In this code, an `Image` object is created and assigned a source. However, before this image can be used in the pattern, you need to ensure it is loaded. Therefore, you must place the rest of the drawing code inside the `Image` object's `onload` event handler. Much like the gradient examples shown earlier, the `Pattern` object that is created with `createPattern()` is then assigned to `fillStyle`. Figure 8-8 in the previous section shows the results.

Summary

In this chapter, you explored how to use the Canvas object as a drawing pad for basic shapes and animations. After a discussion on how to obtain the user agent for Safari on iPhone, you walked through how to draw a rectangle and other shapes, draw an image, and work with properties like color and transparency. You finished by learning about creating image patterns.

9

Special Effects and Advanced Graphics

The strengths of WebKit as a development platform are evident when you begin to explore the capabilities you have with advanced graphics, animation, and other special effects. You can utilize some of the more advanced capabilities of JavaScript, HTML5, and CSS to create cool effects in your Web apps.

I began looking at graphics programming in Chapter 8, "Programming the Canvas." I continue the discussion as I explore how to work with effects and animation.

Gradients

A gradient is a coloring effect you can add to your Web page in which one color gradually changes to another color over the surface of a canvas or other element. *Linear gradients* are applied to a rectangular block surface, whereas *radial gradients* are displayed as circles. You can specify the start and end colors as well as color values in between (known as *color stops*). You can create gradients in both CSS and JavaScript. I'll show you how.

Creating CSS Gradients

You can use the WebKit-supported function `-webkit-gradient()` to create a "virtual" gradient image and define it as an image URL parameter. Here's the syntax for a linear gradient:

```
-webkit-gradient(linear, startPoint, endPoint, from(color), color-stop(percent,
color), to(color))
```

The `linear` parameter defines the type of gradient. The `startPoint` and `endPoint` parameters define the start- and endpoints of the gradient and are typically represented by constants: `left top`, `left bottom`, `right top`, and `right bottom`.

The `from()` and `to()` functions indicate the starting and ending colors, whereas the `color-stop()` function defines a color at a particular point on the gradient.

For example, the following style can be added to a `div` or other block element to create a gradient that starts with light green at the top left and ends with black on the lower left. A color stop of dark green is added to occur at the 50 percent mark. Here's the style:

```
.simpleLinear
{
 width:100px;
 height:100px;
 border:1px solid black;
 background: -webkit-gradient( linear, left top, left bottom, from(#a1f436),
 color-stop(0.5, #668241), to(rgb(0, 0, 0)));
}
```

As you can see, the color can be defined using a hex value or using the `rgb()` function.

The results are shown in Figure 9-1.

Figure 9-1: Linear gradient using CSS

You can add multiple `color-stop()` functions to a gradient to give more complex effect. The following style begins with maroon and ends with white, but it also has green and blue color stops defined at the 30 percent and 80 percent points along the gradient:

```
.complexLinear
{
 width:100px;
 height:100px;
 border:1px solid black;
 background: -webkit-gradient(linear, left top, left bottom, from(#a21c47),
 to(#f9f9f9), color-stop(0.3, #564fb5),color-stop(0.8, #66cc00));
}
```

Figure 9-2 shows the result.

Figure 9-2: Linear gradient with multiple color stops

To create a radial gradient, you use the following syntax:

```
-webkit-gradient(radial, innerCenter, innerRadius, outerCenter, outerRadius,
from(color), color-stop(percent, color), to(color));
```

The `innerCenter` defines the x, y coordinate of the center of the inner circle that starts the gradient. The `innerRadius` defines the width of the radius for the inner circle. The `outerCenter` and `outerRadius` define the same thing for the outer circle that ends the gradient. The `from()`, `to()`, and `color-stop()` functions work the same way as the linear gradient.

The following style example defines a small gradient 10-radial circle starting at (30,30) and ending with a 30-radial circle at (50,50). It has one color stop, midway through the gradient:

```
.radial
{
width:100px;
height:100px;
border:1px solid black;
background: -webkit-gradient(radial, 30 30, 10, 50 50, 30, from(#a1f436),
to(rgba(1,0,0,0)), color-stop(50%, #668241));
}
```

Figure 9-3 displays the result.

Figure 9-3: Radial gradient

You can also combine multiple gradients onto a single surface by listing one after the other. The following style example defines two radial gradients for a 200 • 200 `div`:

```
.radial2
{
width:200px;
height:200px;
border:1px solid black;
background: -webkit-gradient(radial, 20 20, 10, 25 25, 40, from(#a1f436),
to(rgba(1,0,0,0)), color-stop(50%, #668241)),
       -webkit-gradient(radial, 100 100, 20, 20 20, 40, from(#a21c47),
       to(#f9f9f9), color-stop(50%, #66cc00));
}
```

The two orbs are shown in Figure 9-4.

Figure 9-4: Multiple gradients on
one surface

Creating Gradients with JavaScript

If you prefer scripting, you can also create both linear and radial gradients by using the following
methods of the `context` object:

❑　`createLinearGradient(x1,y1,x2,y2)` creates a gradient from the starting point (x1,y1) to
the endpoint (x2,y2).

❑　`createRadialGradient(x1,y1,r1,x2,y2,r2)` creates a gradient circle. The first circle is
based on the x1, y1, and r1 values, and the second circle is based on the x2, y2, and r2 values.

See Chapter 8 for more details on the `canvas` and `context` objects.

Both of these methods return a `canvasGradient` object that can have colors assigned to it with the
`addColorStop(position, color)` method. The `position` argument is a float number between 0.0
and 1.0 that indicates the position of the color in the gradient. The `color` argument is any CSS color.

In the following example, a linear gradient is added to a square box on the canvas. The gradient starts on
the left side, transitions to blue, and ends on the right in red. Here is the code for the entire HTML page:

```
<!DOCTYPE html PUBLIC "-//W3C//DTD XHTML 1.0 Strict//EN"
    "http://www.w3.org/TR/xhtml1/DTD/xhtml1-strict.dtd">
<html xmlns="http://www.w3.org/1999/xhtml">
<head>
<title>Draw Gradient</title>
<meta name="viewport" content="width=320; initial-scale=1.0; maximum-scale=1.0;
```

```
user-scalable=0;">
<script language="javascript" type="text/javascript">
function drawGradient()
{
 var canvas = document.getElementById('myCanvas');
 var context = canvas.getContext('2d');
 var lg = context.createLinearGradient(0,125,250,125);
 context.globalAlpha="0.8";
 lg.addColorStop(0,'white');
 lg.addColorStop(0.75,'blue');
 lg.addColorStop(1,'red');
 context.fillStyle = lg;
 context.strokeStyle="#666666";
 context.lineWidth=".5";
 context.fillRect(10,10,250,250);
 context.strokeRect(10,10,250,250);
}
</script>
</head>
<body onload="drawGradient()">
<canvas id="myCanvas" width="300" height="300" style="position:absolute;
left:0px; top:0px; z-index:1"/>
</body>
</html>
```

The first color stop is set to white, the second to blue, and the third to red. Once you assign these using the `addColorStop()` method, the `lg linearGradient` object is assigned as the `fillStyle` for the context. The `fillRect()` method is then called to paint the block. A gray border is added using the `strokeRect()` method. Figure 9-5 shows the results.

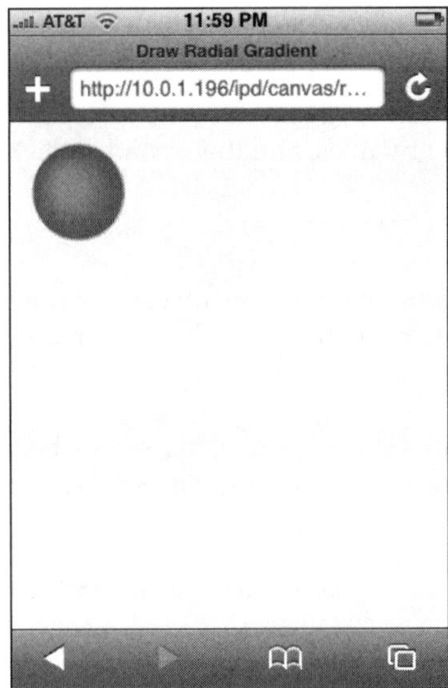

Figure 9-5: A linear gradient can be
set on the canvas

A radial gradient is created by using the `createRadialGradient()` method and then adding color stops at the appropriate position. For example:

```javascript
function drawRadialGradient()
{
  var canvas = document.getElementById('myCanvas');
  var context = canvas.getContext('2d');
  var rg = context.createRadialGradient(45, 45, 10, 52, 50, 35);
  rg.addColorStop(0, '#95b800');
  rg.addColorStop(0.9, '#428800');
  rg.addColorStop(1, 'rgba(220,246,196,0)');
  context.fillStyle = rg;
  context.fillRect(0, 0, 250, 250);
}
```

The `createRadialGradient()` method defines two circles: one with a 10px radius and the second with a 35px radius. Three color stops are added using `addColorStop()`, and then the `rg radialGradient` object is assigned to the `fillStyle` property. See Figure 9-6.

Figure 9-6: Creating a radial gradient

Adding Shadows

Shadow effects are a common visual technique to enhance the look of your page element. Using JavaScript, the `context` object provides four properties that you can use for defining shadows on the canvas. These are as follows:

❑　`shadowColor` defines the CSS color of the shadow.

❑　`shadowBlur` specifies the width of the shadow blur.

❑ shadowOffsetX defines the horizontal offset of the shadow.

❑ shadowOffsetY specifies the vertical offset of the shadow.

Let me show you how to use them. The following code uses these properties to define a blurred shadow for an image:

```
function drawImg(){
  var canvas = document.getElementById('myCanvas');
  var context = canvas.getContext('2d');
  context.shadowColor = "black";
  context.shadowBlur = "10";
  context.shadowOffsetX = "5";
  context.shadowOffsetY = "5";
  var img3 = new Image();
  img3.src = 'images/nola.jpg';
  img3.onload = function()
  {
    context.drawImage(img3, 20, 30);
  }
}
```

The four highlighted lines of code defined a blurred black shadow that is offset 5px from the canvas. Figure 9-7 shows the result.

Figure 9-7: Shadow created using JavaScript

Adding Reflections

If you have used the Cover Flow view in the iTunes app, you have noticed the use of reflections. In fact, image reflections are an increasingly popular effect added to images. Using CSS techniques, you can add reflections to your images or block elements.

To create a reflection, use the `-webkit-box-reflect` property:

```
-webkit-box-reflect : direction offset maskImage|-webkit-gradient;
```

The `direction` parameter specifies the direction of the reflection relative to the object: `above`, `below`, `left`, or `right`. The `offset` parameter defines the offset distance (in pixels or as a percentage) that the reflection should be offset from the block. The `maskImage` parameter specifies a mask image. Or, as shown, you can use a `-webkit-gradient()` function instead. For example, if you wanted to add a bottom reflection to an image using a gradient, you could use the following style rule:

```
<style>

.reflectedImage
{
border:1px solid black;
-webkit-box-reflect:below 3px -webkit-gradient(linear, left top, left bottom,
from(transparent), color-stop(0.5, transparent), to(white));
}


</style>
</head>
<body>

<img src="images/nola.jpg" class="reflectedImage"/>

</body>
```

Notice that the `-webkit-gradient()` function defines a linear gradient as the reflection. It begins fully transparent and ends fully white to blend into the background. Figure 9-8 shows the result.

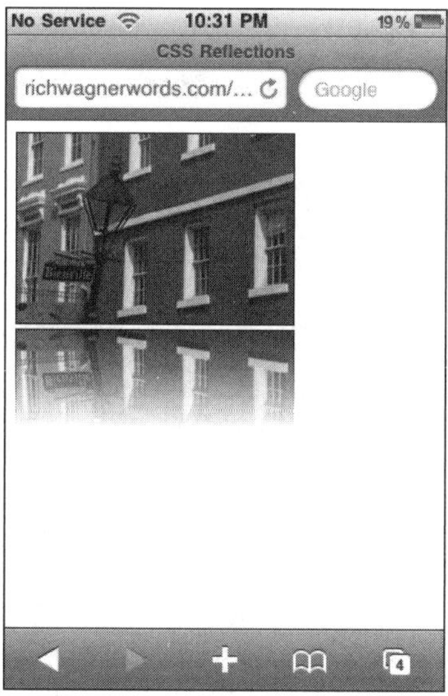

Figure 9-8: Creating a reflection

Reflections on a black background can be especially compelling visually on the iPhone viewport. The following example shows a reflection on black (see Figure 9-9):

```
<!DOCTYPE html PUBLIC "-//W3C//DTD XHTML 1.0 Strict//EN"
      "http://www.w3.org/TR/xhtml1/DTD/xhtml1-strict.dtd">

<html xmlns="http://www.w3.org/1999/xhtml">
<head>
<title>CSS Reflections</title>
<meta name="viewport" content="width=320; initial-scale=1.0; maximum-scale=1.0;
user-scalable=0;">

<style>

  .reflectedImage
  {
    -webkit-box-reflect:below 3px -webkit-gradient(linear, left top, left bottom,
    from(transparent), color-stop(0.5, transparent), to(black));
  }

  body
  {
    background-color:#000000;
  }

</style>
</head>
<body>

<img src="images/nola.jpg" class="reflectedImage"/>

</body>
</html>
```

**Figure 9-9: A reflection can add a
striking effect to your Web app**

If you want to add reflections using JavaScript, I recommend checking out several open source libraries that use canvas programming to create sophisticated charts and image effects, such as reflections. Two particularly noteworthy libraries are the PlotKit and Reflection.js. The PlotKit is a JavaScript Chart Plotting library (available at `www.liquidx.net/plotkit`), and Reflection.js enables you to add reflections to your images (available at `cow.neondragon.net/stuff/reflection`). The Reflection.js library uses canvas to render the reflection but allows you to use it simply by adding a reflect class to an image.

Working with Masks

Masks is a common designer technique used to hide parts of an image. Typically you use a mask to create a blurred border around an image. You can use the CSS property `-webkit-mask-image` to add masks to images. Its syntax is shown here:

```
-webkit-mask-box-image: uri top right bottom left xRepeat yRepeat;
```

The `uri` specifies the mask image, such as the one shown in Figure 9-10. The `top`, `right`, `bottom`, and `left` parameters define the distance the mask is from the edge of the image. The `repeat` parameters specify x, y repeat styles. You can also use the constants: `repeat` (tiled), `stretch`, or `round` (which stretches all parts of the image slightly so that there are no partial tiles at any end).

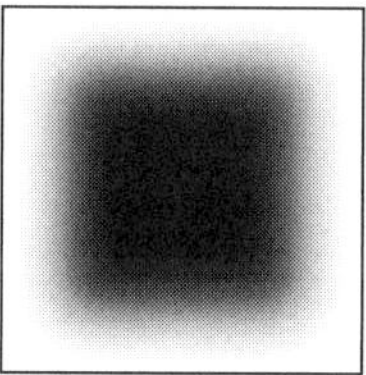

Figure 9-10: Mask image

For example, to create a vignette mask around an image that uses the mask image shown in Figure 9-10, you can use the following style definition:

```
.vignetteMask
{
  -webkit-mask-box-image: url(images/mask.png) 25 stretch;
}
```

The `25` parameter specifies a 25-pixel distance around the full image, and `stretch` stretches the mask (which is smaller than the image it is applied to) to equal the size of the image it is applied to. Figure 9-11 shows the results.

Figure 9-11: Mask applied to an image

You are not limited to using a masked image, however. You can also use a `-webkit-gradient()` function to achieve a faded mask effect. For example, consider this style:

```
.fadeToWhite
{
 -webkit-mask-image:-webkit-gradient(linear, left top, left bottom,
 from(rgba(0,0,0,1)), to(rgba(0,0,0,0)));
}
```

A linear gradient provides a masking effect for the image that uses this style. Notice how the bottom of the image in Figure 9-12 fades into the background.

Or, to achieve a rounded masking effect, you can combine the use of the `-webkit-border-radius` declaration with a `-webkit-mask-image` property. Consider the following:

```
.roundedImage
{
 -webkit-border-radius: 12px;
 -webkit-mask-image:-webkit-gradient(linear, left top, left bottom,
 from(rgba(0,0,0,1)), to(rgba(0,0,0,0)));
}
```

The mask is applied to an image with rounded corners.

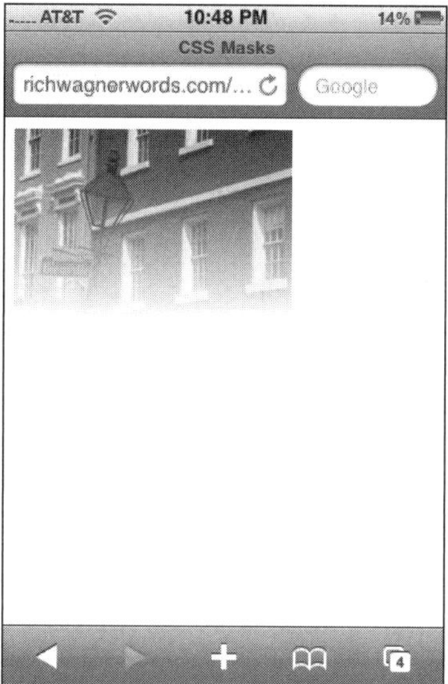

Figure 9-12: Fade effect using a mask

Creating Transform Effects

Back in Chapter 8, I explored how to draw on the canvas using the `context` object. However, I wanted to point out three more methods of the `context` object you can use for transforming the state of a canvas. They are:

- `translate(x, y)` changes the origin coordinates (0, 0) of the canvas.

- `rotate(angle)` rotates the canvas around the current origin by a specified number of radians.

- `scale(x, y)` adjusts the scale of the canvas. The x parameter is a positive number that scales horizontally, whereas the y parameter scales vertically.

The following example uses `translate()` and `scale()` as it draws a circle successive times onto the canvas. Each time these methods are called, their parameters are adjusted:

```
function transform()
{
 var canvas = document.getElementById('myCanvas');
 var context = canvas.getContext('2d');
 var s = 1;
 for (var i = 1; i < 6; i++)
 {
  var t = i * 8;
  context.translate(t, t);
  context.scale(s, s);
```

```
context.fillStyle = "rgba(" + t * 4 + ","+ t * 6 + "," + t * 8 + ", 0.3)";
context.beginPath();
context.arc(50, 50, 40, 0, 2 * pi, false);
context.fill();
s -= 0.05;
  }
}
```

The t variable is 8 times the current iteration of the `for` loop, and then it is used as the parameter for `translate()`. The `scale()` method uses the s variable, which is decremented by 0.05 after each pass. The `fillStyle()` method also uses the t variable to adjust the `rgb` color values for each circle drawn. Figure 9-13 shows the result of the transformation.

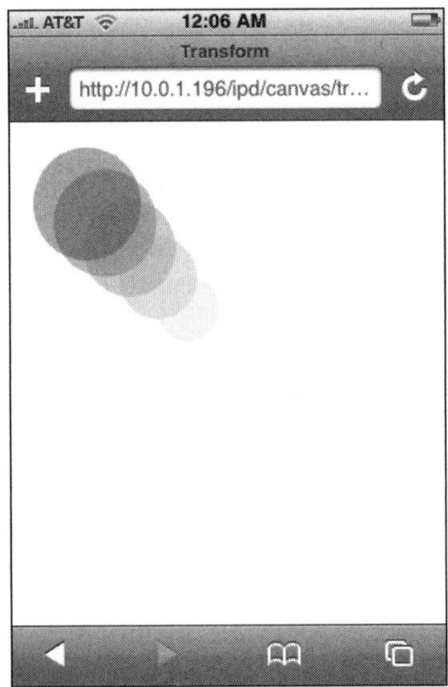

Figure 9-13: A series of transformed circles

The `rotate()` method rotates the canvas based on the specified angle. For example, in the following code, an image is drawn on the canvas three times, and each time the `translate()` and `rotate()` parameter values and the `globalAlpha` property are changed:

```
function rotateImg(){
 var canvas = document.getElementById('myCanvas');
 var context = canvas.getContext('2d');
 context.globalAlpha = "0.5";
   var r = 1;
 var img = new Image();
 img.src = 'images/jared.jpg';
 img.onload = function() {
  for (var i = 1; i < 4;i++) {
   context.translate(50, -15);
   context.rotate(.15 * r);
   context.globalAlpha = i * .33;
```

```
    context.drawImage(img, 20, 20);
     r += 1;
    }
   }
  }
}
```

Figure 9-14 shows the layered result. Note the difference in transparency of the bottommost image to the topmost.

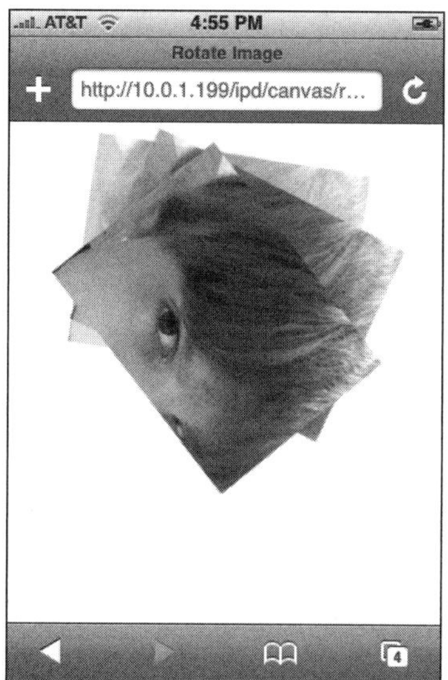

Figure 9-14: Image rotated using rotate()

Note that as you begin to work with more advanced drawings on the canvas, you will need to manage the drawing state. A drawing state includes the current path, the values of the major context properties (such as `fillStyle` and `globalAlpha`), and any transformations (such as rotating) that have been applied. To this end, you can use the `save()` and `restore()` methods. The `save()` method saves a snapshot of the canvas, which can be retrieved later using the `restore()` method. The `save()` and `restore()` methods enable you to return to a default drawing state with minimal additional code and without needing to painstakingly re-create every setting.

Creating Animations

You can use the context drawing capabilities discussed earlier in combination with JavaScript timer routines to create animations on the canvas. On first take, the potential for creating canvas-based animation sounds like a perfect lightweight substitute for Flash for iPhone and iPod touch. For some purposes, it can be ideal. However, any such excitement needs to be kept in reasonable check. Perhaps the chief shortcoming of the canvas drawing in JavaScript is that you need to repaint the entire canvas for each frame of your animation. As a result, complex animations risk becoming jerky on the mobile device. That being said, canvas animation can be a powerful tool to add to your development toolbox.

Like a motion picture or video clip, an animation is a series of frames that, when viewed one after the other, gives the appearance of movement. Therefore, when you code, your job is to show a drawing, clear it, draw the next frame in the series, clear it, and so on until your animation is completed or it loops back to the start. If you are changing any context settings and need to reset them for each new frame, you need to use the `save()` and `restore()` methods.

The following HTML page shows a simple animation program in which a circle moves diagonally from the top-left to the bottom-right part of the canvas:

```
<!DOCTYPE html PUBLIC "-//W3C//DTD XHTML 1.0 Strict//EN"
     "http://www.w3.org/TR/xhtml1/DTD/xhtml1-strict.dtd">

<html xmlns="http://www.w3.org/1999/xhtml">
<head>
<title>Animate</title>
<meta name="viewport" content="width=320; initial-scale=1.0; maximum-scale=1.0;
user-scalable=0;">
<script language="javascript" type="text/javascript">

function init() {
  setInterval( animate, 1 );
}

var p = 0;

function animate(){
  var canvas = document.getElementById('myCanvas');
  var context = canvas.getContext('2d');
  context.globalCompositeOperation = "copy";
  context.fillStyle = "rgba(0,0,255, 0.3)";
  context.beginPath();
  context.arc(50 + p, 50 + p, 30, 0, 360, false);
  context.fill();
  p += 1;
}
</script>
</head>
<body bgcolor="black" onload="init()">
<canvas id="myCanvas" width="300" height="300" style="position:absolute;
left:0px; top:0px"/>
</body>
</html>
```

The `init()` function is called when the document is loaded, which sets off a timer to call `animate()` every 100 milliseconds. The `animate()` function clears the canvas, moves the orientation point, and draws a filled circle. The p variable is then incremented by 1 before repeating.

Figures 9-15 and 9-16 show the start and finish of the animation effect.

Figure 9-15: Start of animation

Figure 9-16: End of animation

Summary

This chapter focused on adding special effects and working with advanced graphic techniques in your iPhone Web apps. You can use both CSS and JavaScript to add gradients. A gradient is a coloring effect in which a one color gradually morphs into another color over the surface of a canvas or other HTML element. There are two types of gradients: linear gradients, which are applied to a rectangular blocks, and radial gradients, which are rendered as circles. To define in CSS, use the `-webkit-gradient()` function to create a "virtual" gradient image and define it as an image URL parameter. To define in JavaScript, use the context object's `createLinearGradient()` or `createRadialGradient()` function.

There are other effects available as well. Shadows can be created with the `shadowColor`, `shadowBlur`, `shadowOffsetX`, and `shadowOffsetY` CSS properties. Reflections are defined using the `-webkit-box-reflect` property. Masks, which hide parts of an image, are defined with the `webkit-mask-image` property.

You can use the canvas drawing in combination with a JavaScript timer to create animations. To do so, you create a drawing, clear it, draw the next frame in the series, clear it, and so on until your animation is completed or it loops back to the start.

10
Integrating with iPhone Services

One of the most intriguing ideas when creating a Web application for iPhone is integrating the application with core mobile services, such as dialing phone numbers or sending e-mails. After all, once you break those inside-the-browser barriers, the application becomes more than just a Web app and extends its functionality across the mobile device.

However, iPhone service integration is a mixed bag; it's a "good news, bad news" situation. On the upside, four of the most important mobile functions (Phone, Mail, SMS Messaging, and Google Maps) are accessible to the developer. On the downside, there is no means of tapping into other core services, such as Calendar, Address Book, Camera, Clock, iPod, and Settings.

To demonstrate the integration with iPhone services, you'll be working with a sample application called iProspector, which is a mocked-up contact management system that emulates the iPhone Contact UI (see Figure 10-1). To create the UI, you will be starting with iUI framework, which is discussed fully in Chapter 3, "Building with Web App Frameworks." However, because it does not provide support for the specific controls needed for the Contact UI, this chapter will show you how to extend iUI as service integration.

Because iPod touch does not provide support for Phone service, any iPhone-specific integration should degrade gracefully when running on iPod touch.

Figure 10-1: Contact UI

Preparing the iProspector Application Shell

Before integrating services and adding custom UI controls for them, you need to prepare the iProspector application shell. The following XHTML document contains a version of the iUI setup for a hierarchical list-based, side-scrolling interface:

```
<!DOCTYPE html PUBLIC "-//W3C//DTD XHTML 1.0 Strict//EN"
        "http://www.w3.org/TR/xhtml1/DTD/xhtml1-strict.dtd">
<html xmlns="http://www.w3.org/1999/xhtml">
<head>
<title>iProspector</title>
<meta name="viewport" content="width=device-width; initial-scale=1.0;
maximum-scale=1.0; user-scalable=0;"/>
<style type="text/css" media="screen">@import "./iui/iui.css";</style>
<style type="text/css" media="screen">@import "./iui/cui.css";</style>
<script type="application/x-javascript" src="./iui/iui.js"></script>
</head>
<body>
    <!--Top iUI toolbar-->
    <div class="toolbar">
        <h1 id="pageTitle"></h1>
        <a id="backButton" class="button" href="#"></a>
        <a class="button" href="#searchForm">Search</a>
    </div>
    <!--Top-level menu-->
```

```html
<!--Customers, Orders, Settings, and About menus not enabled for this sample-->
<ul id="home" title="iProspector" selected="true">
    <li><a href="#leads">Sales Leads</a></li>
    <li><a href="#customers">Customers</a></li>
    <li><a href="#orders">Order Fulfillment</a></li>
    <li><a href="#settings">Settings</a></li>
    <li><a href="#about">About</a></li>
</ul>
<!--Sales Leads menu-->
<ul id="leads" title="Sales Leads">
    <li class="group">A</li>
    <li><a href="#Jack_Armitage">Jack Armitage</a></li>
    <li><a href="#Jason_Armstrong">Jason Armstrong</a></li>
    <li class="group">B</li>
    <li><a href="#Bob_Balancia">Bob Balancia</a></li>
    <li><a href="#Sara_Billingsly">Sara Billingsly</a></li>
    <li><a href="#Uri_Bottle">Uri Bottle</a></li>
    <li><a href="#Larry_Brainlittle">Larry Brainlittle</a></li>
    <li class="group">C</li>
    <li><a href="#Carl_Carlsson">Carl Carlsson</a></li>
    <li><a href="#John_Charleston">John Charleston</a></li>
    <li class="group">D</li>
    <li><a href="#Bill_Drake">Bill Drake</a></li>
    <li><a href="#Randy_Dulois">Randy Dulois</a></li>
</ul>
 <!--Contact panel-->
 <div id="Jack_Armitage" title="Contact" class="panel">
    <h2>This page is intentionally blank.</h2>
</div>
    <!--iUI Search form-->
    <form id="searchForm" class="dialog" action="search.php">
        <fieldset>
            <h1>Contact Search</h1>
            <a class="button leftButton" type="cancel">Cancel</a>
            <a class="button blueButton" type="submit">Search</a>
            <label>Name:</label>
            <input type="text" name="name"/>
            <label>Company:</label>
            <input type="text" name="company"/>
        </fieldset>
    </form>
</body>
</html>
```

In the document head, begin by adding a link to a style sheet named `cui.css`, stored in the same directory as `iui.css`. You'll begin defining `cui.css` shortly.

The iUI framework uses a series of `ul` lists to compose a list-based navigation UI. The `home ul` list provides the top-level menu for the iProspector application (see Figure 10-2). Because you're concerned here with the functionality of working with a specific contact rather than the nuts and bolts of an entire contact management system, the Sales Leads link is the only one defined.

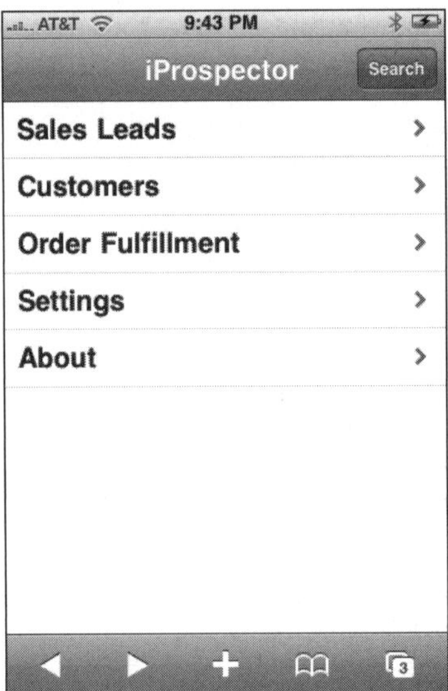

Figure 10-2: iProspector top-level menu

The `leads ul` list provides a canned list of sales leads (see Figure 10-3). Each of the list items contains a link that, in the real world, would be mapped to a unique Contact panel. The `Jack_Armitage` link is connected to the one Contact panel provided in the example document. From a code standpoint, the Contact panel is a `div` element with the `panel` class assigned to it, which displays a generic iPhone-style page (see Figure 10-4).

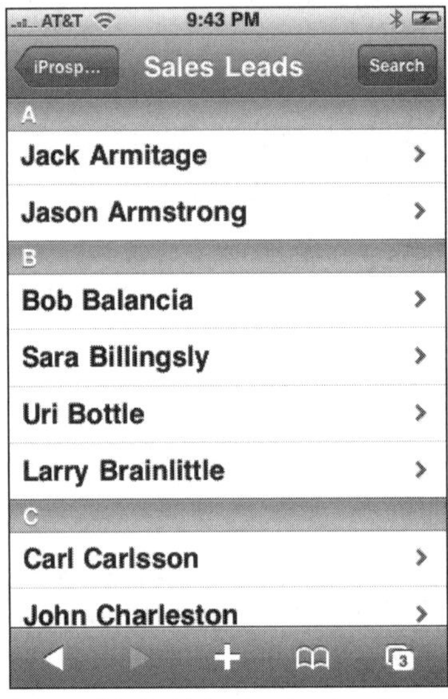

Figure 10-3: List of sales leads

Figure 10-4: Empty Contact panel

Creating the Contact Header

With the application shell functionality in place, the Contact panel is now ready to be filled in. At the top of a typical iPhone Contacts page is a thumbnail image of the contact along with the contact name and company. The HTML document is set up by replacing the dummy h2 text with a `div` element with a `cuiHeader` class that you'll define shortly. Inside of the `div`, three elements are defined, each of which is assigned a `cui` class. Here's the code:

```
<div id="Jack_Armitage" title="Contact" class="panel">
    <div class="cuiHeader">
        <img class="cui" src="jackarmitage.png"/>
        <h1 class="cui">Jack Armitage</h1>
        <h2 class="cui">IBM Corp.</h2>
    </div>
</div>
```

The `img` element will hold the thumbnail image. The `h1` element will contain the name, whereas the `h2` element will show the company.

Creating the cui.css Style Sheet

Next, it is time to create the `cui.css` file (or download it from www.wrox.com). When you use the style conventions originally defined in `iui.css`, four additional rules are defined for the Contact header:

```
.panel h1.cui {
    margin: 5px 0 0px 80px;
    font-size: 20px;
    font-weight: bold;
```

```css
        color: black;
        text-shadow: rgba(255, 255, 255, 0.75) 2px 2px 0;
        top: 5px;
        clear: none;
}
.panel h2.cui {
        margin: 0 0 30px 80px;
        font-size: 14px;
        font-weight: normal;
        color: black;
        text-shadow: rgba(255, 255, 255, 0.75) 2px 2px 0;
        top: 43px;
        clear: none;
}
.panel img.cui {
        margin: 0px 15px 5px 0px;
        border: 1px solid #666666;
        float: left;
        -webkit-border-radius: 5px;
}
.panel > div.cuiHeader {
        position: relative;
        margin-bottom: 0px 0px 10px 14px;
}
```

The first three rules position the h1, h2, and img elements in the appropriate location. The final rule adds spacing between the header panel and the rest of the page. Figure 10-5 shows the current state of the Contact panel.

With all the preparatory UI in place, you can begin to add the service integration.

Figure 10-5: Adding the Contact header

Making Phone Calls from Your Application

You can make a phone call from your application through a special telephone link. A telephone link is specified through the `tel:` protocol. The basic syntax is this:

```
<a href="tel:1-507-555-5555">1-507-555-5555</a>
```

When a user clicks the link, the phone does not automatically dial. Instead, iPhone displays a confirmation box (see Figure 10-6) that allows the user to click Call or Cancel.

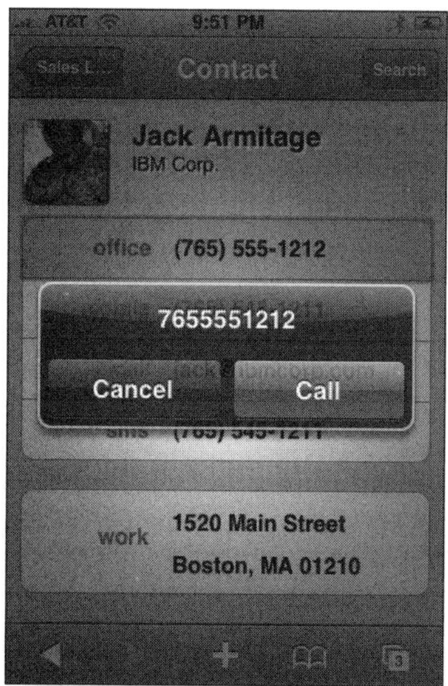

Figure 10-6: User needs to confirm a telephone link before a call is initiated.

Telephone links can go beyond ordinary numbers. iPhone provides partial support for the RFC 2086 protocol (`www.ietf.org/rfc/rfc2806.txt`), which enables you to develop some sophisticated telephone-based URLs. For example, the following link calls the U.S. Postal Service, pauses for 2 seconds, and then presses 2 to get a Spanish version:

```
<a href="tel:+1-800-ASK-USPS;pp2">USPS (Espanol)</a>
```

Note that p creates a 1-second pause, so pp will cause a 2-second pause before continuing. Safari on iPhone will also automatically create telephone links for you in your pages. Any number that takes the form of a phone is displayed as a link. Therefore, if you ever have a situation in which you do not want to link a telephone number (or a number that could be interpreted as a phone number), add the `format-detection` <meta> tag to the document head:

```
<meta name = "format-detection" content = "telephone=no">
```

Creating Service Links

In adding this telephone link functionality into iProspector, you want to emulate the telephone links inside the iPhone Contact UI. To do so, begin by adding a `fieldset` in `prospector.html` and enclosing two `row` `div` elements inside of it. Inside of the `div` elements, add a label and a link. Here's the code:

```
<fieldset>
    <div class="row">
        <label class="cui">office</label>
        <a class="cuiServiceLink" target="_self" href="tel:(765) 555-1212"
        >(765) 555-1212</a>
    </div>
    <div class="row">
        <label class="cui">mobile</label>
        <a class="cuiServiceLink" target="_self" href="tel:(765) 545-1211"
        >(765) 545-1211</a>
    </div>
</fieldset>
```

The a links, which are referred to as service links in this book, are assigned a `cuiServiceLink` class and use the `tel:` protocol in the `href` value. The `target="_self"` attribute is needed to override default iUI behavior, which would prevent the link from calling the Phone application. Finally, the `label` is assigned a `cui` class.

The `fieldset` and `row` class styling are already defined in the `iui.css`. However, several additional styles need to be defined inside the `cui.css` file. First, styles need to be defined for the labels and service links. Second, a set of styles needs to be added to emulate the push-down effect of the services link when a user presses it. The rules are shown in the following code:

```css
.row > label.cui {
    position: absolute;
    margin: 0 0 0 14px;
    line-height: 42px;
    font-weight: bold;
    color: #7388a5;
}
.cuiServiceLink {
    display: block;
    margin: 0;
    border: none;
    padding: 12px 10px 0 80px;
    text-align: left;
    font-weight: bold;
    text-decoration: inherit;
    height: 42px;
    color: inherit;
    box-sizing: border-box;
}
.row[cuiSelected] {
    position: relative;
    min-height: 42px;
    border-bottom: 1px solid #999999;
    -webkit-border-radius: 0;
```

```
        text-align: right;
        background-color: #194fdb !important;
        color: #FFFFFF !important;
    }
    .row[cuiSelected] > label.cui  {
        position: absolute;
        margin: 0 0 0 14px;
        line-height: 42px;
        font-weight: bold;
        color: #FFFFFF;
    }
    fieldset > .row[cuiSelected]:last-child  {
        border-bottom: none !important;
    }
```

The bottom three rules are used to change the `row` and `label` styling when the `row div` has a `cuiSelected` attribute set to `true`. (The element's background becomes blue, and the label font is set to white.)

Although the styles are now ready for these elements, the service links are not yet functional within the iUI framework. By default, iUI intercepts all link click events inside of `iui.js` to change a link's selection state and to disable the default action of a link. Therefore, you need to add a handler for service link buttons coming through this routine. Here's the modified version of the `addEventListener("click", function(event))` handler:

```
addEventListener("click", function(event)
{
    var link = findParent(event.target, "a");
    if (link)
    {
        function unselect() { link.removeAttribute("selected"); }
        if (link.href && link.hash && link.hash != "#")
        {
            link.setAttribute("selected", "true");
            iui.showPage($(link.hash.substr(1)));
            setTimeout(unselect, 500);
        }
        // Begin cui insertion
        else if ( link.getAttribute("class") == "cuiServiceLink" )
          {
            var curRow = findParent( link, "div" );
            curRow.setAttribute("cuiSelected", "true");
                          setTimeout(function() {
                              curRow.removeAttribute("cuiSelected");
                  }, 500);
            return;
          }
        // End cui insertion
        else if (link == $("backButton"))
            history.back();
        else if (link.getAttribute("type") == "submit")
            submitForm(findParent(link, "form"));
        else if (link.getAttribute("type") == "cancel")
```

```
            cancelDialog(findParent(link, "form"));
        else if (link.target == "_replace")
        {
            link.setAttribute("selected", "progress");
            iui.showPageByHref(link.href, null, null, link, unselect);
        }
        else if (!link.target)
        {
            link.setAttribute("selected", "progress");
            iui.showPageByHref(link.href, null, null, null, unselect);
        }
        else
            return;
        event.preventDefault();
    }
}, true);
```

The first `else if` conditional block is inserted to check for all links that have a class of `cuiSer-viceLink`. If one is discovered, the parent `div` is retrieved and its instance assigned to `curRow`. A `cui-Selected` attribute is then added to the `curRow` and removed. When paired with the styles set up in `cui.css`, this code changes the colors of the service link's parent `div` for 500 milliseconds; then it sets them back to normal. The visual effect simulates, as much as possible, the default behavior of iPhone. Finally, a `return` statement is added at the end of the block to ensure that the `preventDefault()` command is not called (which would prevent the services link from working correctly).

The telephone links of the Contact panel, shown in Figure 10-7, are now styled and fully functional.

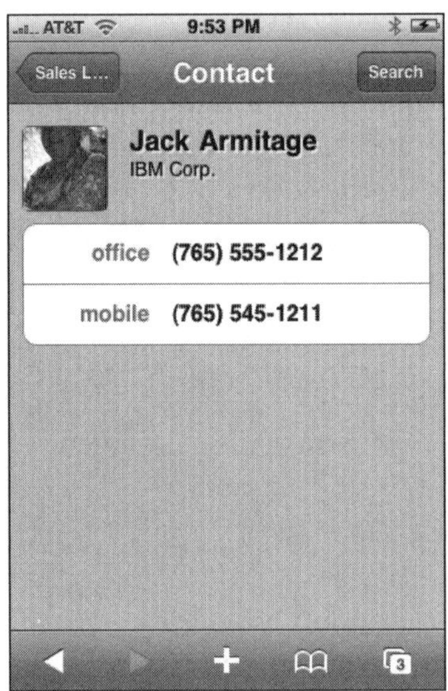

Figure 10-7: Telephone links added
to the Contact panel

Sending E-Mails

E-mails can also be sent from your application through links using the familiar `mailto:` protocol, as shown in the following example:

```
<a href="mailto:jack@ibmcorp.com">Jack Armitage</a>
```

When this link is clicked by the user, Mail opens and a new message window is displayed, as shown in Figure 10-8. The user can then fill out the subject and body of the message and send it. As you would expect, you cannot automatically send an e-mail message using the `mailto:` protocol without user intervention. The `mailto:` protocol always takes the user to a new message window.

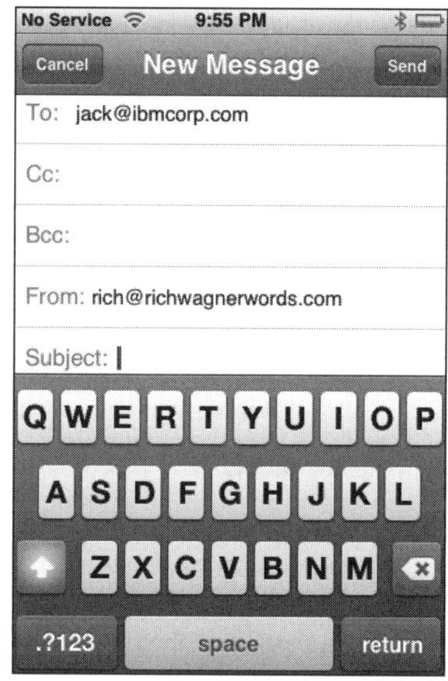

Figure 10-8: Sending a mail message from an application.

Following the `mailto:` protocol, you can also include parameters to specify the subject, cc address, bcc address, and message body. Table 10-1 lists these options.

Table 10-1: Optional mailto: Attributes

Option	Syntax
Multiple recipients	, (comma separating e-mail addresses)
Message subject	subject=Subject Text
cc recipients	cc=name@address.com
bcc recipients	bcc=name@address.com
Message text	body=Message text

Per HTTP conventions, precede the initial parameter with a ? (such as ?subject=), and precede any additional parameters with an &.

The `mailto:` protocol normally allows line breaks in the `body` attribute value using `%0A` for a line break and `%0A%0A` for a line break followed by a blank line. However, iPhone ignores the `%0A` codes and puts all the text on one line.

As a work-around, iPhone enables you to embed HTML in your message body, thereby enabling you to add `br` tags for line breaks and even other tags (such as `strong`) for formatting.

When you combine several parameters, the following element provides everything a user needs to send a reminder message:

```
<a  class="cuiServiceButton" target="_self" onclick="return
(navigator.userAgent.indexOf('iPhone') != -1)"href="mailto:jack@ibmcorp.com?
subject=Meeting&body=Dear Jack,<br/>I look forward to our upcoming meeting
together <strong>this Friday at 8am.</strong><br/>Sincerely,<br/>Jason
Malone&cc=jason@iphogcorp.com">Email Reminder</a>
```

As Figure 10-9 shows, all the user needs to do is press the Send button.

Figure 10-9: Populating an e-mail message with data from an application.

Adding an e-mail link to the iProspector application is straightforward. Because the look and functionality of an e-mail link are identical to those of telephone links in the native iPhone Contact UI, you can piggyback on top of the styles and functionality you defined earlier in this chapter. With that in mind, add an e-mail link just under the two telephone links inside of the same `fieldset`:

```
<fieldset>
    <div class="row">
        <label class="cui">office</label>
```

```
        <a class="cuiServiceLink" target="_self" href="tel:(765) 555-1212"
        >
        (765) 555-1212</a>
    </div>
    <div class="row">
        <label class="cui">mobile</label>
        <a class="cuiServiceLink" target="_self" href="tel:(765) 545-1211"
        >
        (765) 545-1211</a>
    </div>
    <div class="row">
        <label class="cui">email</label>
        <a class="cuiServiceLink" target="_self"
        href="mailto:jack@ibmcorp.com">
        jack@ibmcorp.com</a>
    </div>
</fieldset>
```

Sending SMS Messages

Much like e-mail message integration, you can send SMS messages using the sms protocol. For example, the following code launches the SMS application, addressing a text message to 765-545-1211:

```
<a href="sms:765-545-1211">(765) 545-1211</a>
```

As Figure 10-10 shows, you can then enter the message using the keyboard and send it when finished.

Alternatively, if you just want to launch the SMS app without a specific number to text, use a blank sms:

```
<a href="sms:">Send SMS Message</a>
```

For the iProspector app, you can add SMS message support much like the e-mail button, as shown in the code highlighted here:

```
<fieldset>
    <div class="row">
        <label class="cui">office</label>
        <a class="cuiServiceLink" target="_self" href="tel:(765) 555-1212"
        >
        (765) 555-1212</a>
    </div>
    <div class="row">
        <label class="cui">mobile</label>
      <a class="cuiServiceLink" target="_self" href="tel:(765) 545-1211"
        >
        (765) 545-1211</a>
    </div>
    <div class="row">
        <label class="cui">email</label>
        <a class="cuiServiceLink" target="_self"
        href="mailto:jack@ibmcorp.com">
        jack@ibmcorp.com</a>
    </div>
```

```
            <div class="row">
              <label class="cui">sms</label>
                <a class="cuiServiceLink" target="_self" href="sms:765-545-1211">
                (765) 545-1211</a>
            </div>
          </fieldset>
```

Figure 10-10: SMS application can be launched from a Web app using the `sms` prototcol.

If you try to click an SMS link on an iPod touch, the device ignores the request, since iPod touch does not support SMS messaging.

Pointing on Google Maps

Although Google Maps does not have its own custom `href` protocol, Safari on iPhone is smart enough to reroute any request to `maps.google.com` to the built-in Maps application rather than going to the public Google Web site. (On iPod touch, Safari links directly to the public Google Web site.) As a result, you can create a link to specify either a specific location or driving directions between two geographical points.

You cannot specify whether to display the map in Map or Satellite view. The location you specify will be displayed in the last selected view of the user.

Keep in mind the basic syntax conventions when composing a Google Maps URL:

❑   For normal destinations, start with the `q=` parameter, and then type the location as you would a normal address, substituting + signs for blank spaces.

❑   For clarity, include commas between address fields.

Here's a basic URL to find a location based on city and state:

```
<a href="http://maps.google.com/maps?q=Boston,+MA">Boston</a>
```

Here's the syntax used for a full street address:

```
<a href="http://maps.google.com/maps?q=1000+Massachusetts+Ave,+Boston,+MA">Jack
Armitage's Office</a>
```

When the address shown previously is located in Google Maps, the marker is generically labeled `1000 Massachusetts Ave Boston MA`. However, you can specify a custom label by appending the URL with `+(Label+Text)`, as shown in the following example:

```
<a  href="http://maps.google.com/maps?q=1000+Massachusetts+Ave,+Boston,+MA+
(Jack+Armitage's+Office)">Jack Armitage's Office</a>
```

Figure 10-11 shows the custom label in Google Maps.

Figure 10-11: Customizing the Google Maps label

You can also specify a location using latitude and longitude coordinates:

```
<a  href="http://maps.google.com/maps?q=52.123N,2.456W">Jack's Summer Retreat</a>
```

To get directions, use the `saddr=` parameter to indicate the starting address and the `daddr=` parameter to specify the destination address, as shown in the following example:

```
<a href="http://maps.google.com/maps?saddr=Holden+MA&daddr=1000+Massachusetts+Ave,
+Boston,+MA">Directions To Office</a>
```

Figure 10-12 displays the map view when this link is clicked.

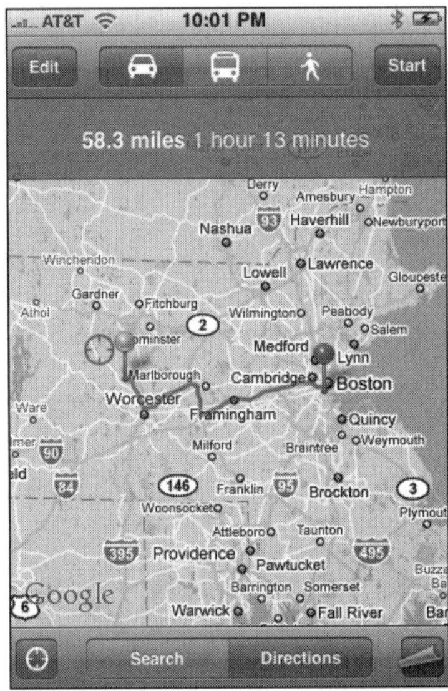

Figure 10-12: Programming driving directions

The Google Maps public Web site has an extensive set of parameters. However, except where noted previously, none of these are supported at this time. You cannot, for example, use the t= parameter to specify the Satellite map, the z= parameter to indicate the map zoom level, or even layer=t to turn on the Traffic display. The user needs to perform those steps interactively.

To include Google Maps integration with iProspector, you need to add two new capabilities to its Contact panel. First, you need to display multiline, read-only address information in its own box. Second, you need to create a new action button style to emulate the button functionality of the native iPhone Contact UI.

Creating a Contacts Address Box

To define an address box, define a `div` with a new style named `rowCuiAddressBox`. Inside of it, add a `cui label` and then `cui p` elements for each line of the address:

```
<fieldset>
    <div class="rowCuiAddressBox">
            <label class="cui">work</label>
            <p class="cui">1520 Main Street</p>
            <p class="cui">Boston, MA 01210</p>
    </div>
</fieldset>
```

Next, going back to `cui.css`, you need to define four new styles:

```css
.rowCuiAddressBox   {
    position: relative;
    min-height: 24px;
    border-bottom: 1px solid #999999;
    -webkit-border-radius: 0;
    text-align: left;
}
.rowCuiAddressBox > p.cui {
    box-sizing: border-box;
    margin: 0;
    border: none;
    text-align: left;
    padding: 2px 10px 0 80px;
    height: 30px;
    background: none;
    font-weight: bold;
}
fieldset > .rowCuiAddressBox:first-child  {
     padding-top: 12px;
    border-bottom: none !important;
}
fieldset > .rowCuiAddressBox:last-child  {
    min-height: 25px;
    text-align: left;
    border-bottom: none !important;
}
```

The `:first-child` and `:last-child` styles ensure proper padding and sizing of the contents of the box.

To style the address box label, you need to add one additional selector onto the previously defined `.row > label.cui` rule:

```css
.row > label.cui,  .rowCuiAddressBox > label.cui  {
    position: absolute;
    margin: 0 0 0 14px;
    line-height: 42px;
    font-weight: bold;
    color: #7388a5;
}
```

The display-only address box is now ready.

Creating Service Buttons

Two new links are needed to add Google Maps integration. One link will display a map of the contact, and a second will provide driving directions. Here is the `fieldset` definition:

```html
<fieldset>
    <div class="row">
```

```
                <a  class="cuiServiceButton" target="_self"
    href="http://maps.google.com/maps?q=1000+Massachusetts+Ave,+Boston,+MA">Map To
    Office</a>
            </div>
            <div class="row">
                <a class="cuiServiceButton" target="_self"
    href="http://maps.google.com/maps?saddr=Holden+MA&daddr=1000+Massachusetts+Ave,
    +Boston,+MA">Directions To Office</a>
            </div>
        </fieldset>
```

These two links are assigned to the `cuiServiceButton` class. The first link displays a map of the
specified address in Boston, whereas the second link provides driving directions between Holden,
Massachusetts and the Boston address. Once again, to get around the way iUI handles events in
`iui.jss`, you need to specify the `target="_self"` parameter.

Back over in `cui.css`, one new style needs to be added:

```
.cuiServiceButton {
    display: block;
    margin: 0;
    border: none;
    padding: 12px 10px 0 0px;
    text-align: center;
    font-weight: bold;
    text-decoration: inherit;
    height: 42px;
    color: #7388a5;
    box-sizing: border-box;
}
```

This style emulates the look of the action buttons (centered blue text, and so on) in the native iPhone
Contact UI.

There is one final tweak that needs to be made to `iui.jss` before the `cuiServiceButton` links work
as expected. If you recall, an `else if` condition is added to trap for service links inside the `addEvent-
Listener("click", event(function))` function. You need to add an additional test so that both
`cuiServiceLink` and `cuiServiceButton` classes are evaluated. To do so, modify the line of code as
specified here:

```
else if ( (link.getAttribute("class") == "cuiServiceLink" ) ||
( link.getAttribute("class") == "cuiServiceButton") )
```

Now that the `cuiServiceButton` link class is ready to go, you need to add one last button to the iProspector Contact panel to finish it off — a services button that automatically composes a reminder e-mail
to the Contact. The following HTML code combines `mailto:` link functionality and the `cuiService-
Button` style:

```
    <fieldset>
        <div class="row">
            <a  class="cuiServiceButton" target="_self" href="mailto:
    jack@ibmcorp.com?subject=Meeting&body=Dear Jack, I look forward to our upcoming
```

```
meeting together this Friday at 8am. Sincerely, Jason Malone&cc=
jason@iphogcorp.com">Email Reminder</a>
            </div>
          </fieldset>
```

Figure 10-13 shows the display of these `cuiServiceButton` links inside of iProspector.

The iProspector Contact panel is now fully enabled to emulate both the look and functionality of the built-in iPhone Contact UI.

Listing 10-1 displays the `prospector.html` file, Listing 10-2 displays the `cui.css` file, and Listing 10-3 displays the modified function block inside of `iui.jss`.

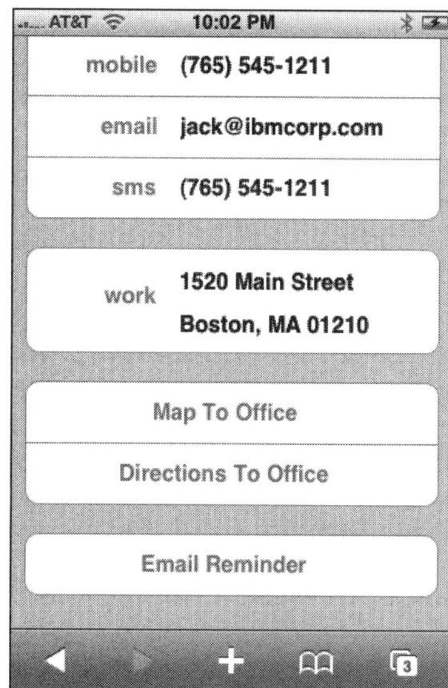

Figure 10-13: Enabled Contact buttons
that integrate with Google Maps and Mail

Listing 10-1: prospector.html

```
<!DOCTYPE html PUBLIC "-//W3C//DTD XHTML 1.0 Strict//EN"
         "http://www.w3.org/TR/xhtml1/DTD/xhtml1-strict.dtd">
<html xmlns="http://www.w3.org/1999/xhtml">
<head>
<title>iProspector</title>
<meta name="viewport" content="width=320; initial-scale=1.0;
maximum-scale=1.0; user-scalable=0;"/>
<style type="text/css" media="screen">@import "./iui/iui.css";</style>
<style type="text/css" media="screen">@import "./iui/cui.css";</style>
<script type="application/x-javascript" src="./iui/iui.js"></script>
</head>
```

Continued

Listing 10-1: prospector.html *(continued)*

```html
<body>
            <!--Top iUI toolbar-->
    <div class="toolbar">
        <h1 id="pageTitle"></h1>
        <a id="backButton" class="button" href="#"></a>
        <a class="button" href="#searchForm">Search</a>
    </div>
    <!--Top-level menu-->
    <!--Customers, Orders, Settings, and About menus not enabled for
    this sample-->
    <ul id="home" title="iProspector" selected="true">
        <li><a href="#leads">Sales Leads</a></li>
        <li><a href="#customers">Customers</a></li>
        <li><a href="#orders">Order Fulfillment</a></li>
        <li><a href="#settings">Settings</a></li>
        <li><a href="#about">About</a></li>
    </ul>
    <!--Sales Leads menu-->
    <ul id="leads" title="Sales Leads">
        <li class="group">A</li>
        <li><a href="#Jack_Armitage">Jack Armitage</a></li>
        <li><a href="#Jason_Armstrong">Jason Armstrong</a></li>
        <li class="group">B</li>
        <li><a href="#Bob_Balancia">Bob Balancia</a></li>
        <li><a href="#Sara_Billingsly">Sara Billingsly</a></li>
        <li><a href="#Uri_Bottle">Uri Bottle</a></li>
        <li><a href="#Larry_Brainlittle">Larry Brainlittle</a></li>
        <li class="group">C</li>
        <li><a href="#Carl_Carlsson">Carl Carlsson</a></li>
        <li><a href="#John_Charleston">John Charleston</a></li>
        <li class="group">D</li>
        <li><a href="#Bill_Drake">Bill Drake</a></li>
        <li><a href="#Randy_Dulois">Randy Dulois</a></li>
    </ul>
        <!--Contact panel-->
    <div id="Jack_Armitage" title="Contact" class="panel">
                <div class="cuiHeader">
                        <img class="cui" src="jackarmitage.png"/>
            <h1 class="cui">Jack Armitage</h1>
            <h2 class="cui">IBM Corp.</h2>
        </div>
        <fieldset>
            <div class="row">
                <label class="cui">office</label>
                <a class="cuiServiceLink" target="_self" href=
                "tel:(765) 555-1212"
                >(765) 555-1212</a>
            </div>
            <div class="row">
                <label class="cui">mobile</label>
                <a class="cuiServiceLink" target="_self" href="tel:(765)
                545-1211"
```

```
            >(765) 545-1211</a>
        </div>
        <div class="row">
            <label class="cui">email</label>
            <a class="cuiServiceLink" target="_self"
            href="mailto:jack@ibmcorp.com"
            >jack@ibmcorp.com</a>
        </div>
         <div class="row">
            <label class="cui">sms</label>
            <a class="cuiServiceLink" target="_self" href="sms:
            765-545-1211">(765) 545-1211</a>
        </div>
    </fieldset>
    <fieldset>
        <div class="rowCuiAddressBox">
                    <label class="cui">work</label>
                    <p class="cui">1520 Main Street</p>
                    <p class="cui">Boston, MA 01210</p>
        </div>
    </fieldset>
    <fieldset>
        <div class="row">
            <a  class="cuiServiceButton" target="_self"
href="http://maps.google.com/maps?q=1000+Massachusetts+Ave,
+Boston,+MA">Map To Office</a>
        </div>
        <div class="row">
            <a  class="cuiServiceButton" target="_self"
href="http://maps.google.com/maps?saddr=Holden+MA&daddr=1000+
Massachusetts+Ave,+Boston,+MA">Directions To Office</a>
        </div>
    </fieldset>
    <fieldset>
        <div class="row">
            <a  class="cuiServiceButton" target="_self"
            onclick="return
            (navigator.userAgent.indexOf('iPhone') != -1)
            "href="mailto:jack@ibmcorp.com?subject=
            Meeting&body=Dear Jack,<br/>I look forward to our
            upcoming meeting together <strong>this Friday at
            8am.</strong><br/>Sincerely,<br/>Jason
            Malone&cc=jason@iphogcorp.com">Email Reminder</a>
        </div>
    </fieldset>
</div>
        <!--iUI Search form-->
<form id="searchForm" class="dialog" action="search.php">
    <fieldset>
        <h1>Contact Search</h1>
        <a class="button leftButton" type="cancel">Cancel</a>
        <a class="button blueButton" type="submit">Search</a>
        <label>Name:</label>
        <input type="text" name="name"/>
```

Continued

Listing 10-1: prospector.html *(continued)*

```html
                <label>Company:</label>
                <input type="text" name="company"/>
            </fieldset>
        </form>
    </body>
</html>
```

Listing 10-2: cui.css

```css
/* cui Contacts Extension to Joe Hewitt's iUI */
/* Contact Header */
.panel h1.cui {
    margin: 5px 0 0px 80px;
    font-size: 20px;
    font-weight: bold;
    color: black;
    text-shadow: rgba(255, 255, 255, 0.75) 2px 2px 0;
     top: 5px;
     clear: none;
}
.panel h2.cui {
    margin: 0 0 30px 80px;
    font-size: 14px;
    font-weight: normal;
    color: black;
    text-shadow: rgba(255, 255, 255, 0.75) 2px 2px 0;
     top: 43px;
     clear: none;
}
.panel img.cui {
    margin: 0px 15px 5px 0px;
    border: 1px solid #666666;
    float: left;
    -webkit-border-radius: 5px;
}
.panel > div.cuiHeader {
    position: relative;
    margin-bottom: 0px 0px 10px 14px;
}
/* Contact Fields */
.row > label.cui, .rowCuiAddressBox > label.cui  {
    position: absolute;
    margin: 0 0 0 14px;
    line-height: 42px;
    font-weight: bold;
    color: #7388a5;
}
.cuiServiceLink {
    display: block;
    margin: 0;
    border: none;
```

```css
    padding: 12px 10px 0 80px;
      text-align: left;
    font-weight: bold;
    text-decoration: inherit;
    height: 42px;
    color: inherit;
     box-sizing: border-box;
}
.cuiServiceButton {
    display: block;
    margin: 0;
    border: none;
    padding: 12px 10px 0 0px;
      text-align: center;
    font-weight: bold;
    text-decoration: inherit;
    height: 42px;
    color: #7388a5;
     box-sizing: border-box;
}
a[cuiSelected], a:active {
    background-color: #194fdb !important;
    color: #FFFFFF !important;
}
.row[cuiSelected]  {
    position: relative;
    min-height: 42px;
    border-bottom: 1px solid #999999;
    -webkit-border-radius: 0;
    text-align: right;
    background-color: #194fdb !important;
    color: #FFFFFF !important;
}
.row[cuiSelected] > label.cui  {
    position: absolute;
    margin: 0 0 0 14px;
    line-height: 42px;
    font-weight: bold;
    color: #FFFFFF;
}
fieldset > .row[cuiSelected]:last-child  {
    border-bottom: none !important;
}
/* Contact Address Box (Display-only)  */
.rowCuiAddressBox    {
    position: relative;
    min-height: 24px;
    border-bottom: 1px solid #999999;
    -webkit-border-radius: 0;
    text-align: left;
}
.rowCuiAddressBox > p.cui {
    box-sizing: border-box;
    margin: 0;
```

Continued

Listing 10-2: cui.css *(continued)*

```
        border: none;
        text-align: left;
        padding: 2px 10px 0 80px;
        height: 30px;
        background: none;
        font-weight: bold;
    }
fieldset > .rowCuiAddressBox:first-child  {
        padding-top: 12px;
        border-bottom: none !important;
    }
fieldset > .rowCuiAddressBox:last-child  {
        min-height: 25px;
        text-align: left;
        border-bottom: none !important;
    }
```

Listing 10-3: Modified portion of iui.js

```
addEventListener("click", function(event)
{
    var link = findParent(event.target, "a");
    if (link)
    {
        function unselect() { link.removeAttribute("selected"); }
        if (link.href && link.hash && link.hash != "#")
        {
            link.setAttribute("selected", "true");
            iui.showPage($(link.hash.substr(1)));
            setTimeout(unselect, 500);
        }
        // Begin cui insertion
        else if ( (link.getAttribute("class") == "cuiServiceLink" )  || ( link
.getAttribute("class") == "cuiServiceButton") )
            {
                var curRow = findParent( link, "div" );
                curRow.setAttribute("cuiSelected", "true");
                        setTimeout(function() {
                            curRow.removeAttribute("cuiSelected");
                }, 500);
                return;
            }
         // End cui insertion
        else if (link == $("backButton"))
            history.back();
        else if (link.getAttribute("type") == "submit")
            submitForm(findParent(link, "form"));
        else if (link.getAttribute("type") == "cancel")
            cancelDialog(findParent(link, "form"));
        else if (link.target == "_replace")
        {
```

```
            link.setAttribute("selected", "progress");
            iui.showPageByHref(link.href, null, null, link, unselect);
        }
        else if (!link.target)
        {
            link.setAttribute("selected", "progress");
            iui.showPageByHref(link.href, null, null, null, unselect);
        }
        else
            return;
    }
}, true);
```

Summary

One of the most compelling features of a mobile application — either Web-based or native — is the ability to integrate with core iPhone services. In this chapter, I showed you how you can integrate your iPhone Web applications with four of the most important mobile functions: Phone, Mail, SMS Messaging, and Google Maps. As I did so, I walked you through how you might utilize those services in a contact management app.

11

Offline Applications

In the past, one of the key differences between native iPhone apps and Web apps was the ability that native apps had to work with local and remote data, whereas iPhone Web apps were limited to working only when a live connection was available. However, Safari on iPhone has embraced support for HTML 5's offline capabilities, enabling you to create Web apps that work even when the user has no access to the Internet.

In this chapter, I'll walk you through these offline capabilities.

The HTML 5 Offline Application Cache

Safari on iPhone takes advantage of HTML 5's offline application cache to enable you to pull down remote files from a Web server and store them in a local cache.

In this way, when the device is not connected to the Internet, either through 3G or Wifi access, users can continue to work with your Web app, just in offline mode.

You can include any file in the manifest that can be displayed locally without server-side processing — images (JPG, PNG, and GIF), HTML files, CSS style sheets, and JavaScript scripts.

Safari then attempts to download files in the manifest. If successful, Safari looks for these files in the cache before going to the server. However, in the event of a missing file, incorrect URL, or other error, the update process fails and no further files are downloaded. Then, the next time the manifest is loaded, Safari attempts to download all files once again.

Once Safari downloads the files in a manifest file, the cache is only updated in the future if the manifest file changes. (Note that the content of the manifest file is what is evaluated to determine whether to update the cache, not its last saved date or any other file attribute.) However, if you want to force an update, you can do so programmatically through JavaScript.

Creating a Manifest File

To enable the offline cache, you need to create a manifest file that lists each of the files you want to have available offline. The manifest file is an ordinary text file, without HTML or XML markup, that includes two parts:

❑ **Declaration:** The manifest is declared by typing the following on the first line of the file:

```
CACHE MANIFEST
```

❑ **URL listings:** The lines that follow list the URLs for each file that you want to cache. The paths must be relative to the path of the manifest file.

You can also add comments to the file by adding a # to the start of each line.

Here's a sample manifest file used for a small Web app. Note that this app caches a subset of iUI resources as part of the local cache:

```
CACHE MANIFEST

# images
jackarmitage.png

# in-use iUI files
../iui/iui.css
../iui/cui.css
../iui/iui.js
../iui/whiteButton.png
../iui/toolButton.png
../iui/toolbar.png
../iui/toggleOn.png
../iui/toggle.png
../iui/thumb.png
../iui/selection.png
../iui/prev.png
../iui/pinstripes.png
../iui/next.png
../iui/loading.gif
../iui/listGroup.png
../iui/listArrowSel.png
../iui/listArrow.png
../iui/grayRow.png
../iui/grayButton.png
../iui/cancel.png
../iui/blueButton.png
../iui/blackToolButton.png
../iui/blackToolbar.png
../iui/blackButton.png
../iui/backButton.png
```

Once the manifest file is created, save it with a `.manifest` extension. For example, in my example, I named it `prospector.manifest`.

For offline cache to work correctly, you need to be sure that your Web server serves up the manifest file correctly. Because HTML 5 offline cache is still "cutting-edge" technology, many ISPs do not provide built-in support for it. Therefore, check to make sure your server assigns the MIME type `text/cache-manifest` to the `manifest` extension. In my case, I had to add it as a custom MIME type.

Referencing the Manifest File

After you have created the manifest file and uploaded it to your server, you need to link it to your Web app. To do so, add the `manifest` attribute to the root `html` tag of your Web file:

```
<html manifest="prospector.manifest" xmlns="http://www.w3.org/1999/xhtml">
```

In this case, the `prosector.manifest` file is in the same directory as the `index.html` file.

Programmatically Controlling the Cache

You have access to the cache using the JavaScript object `window.applicationCache` from inside your Web app. To force a cache update, you can use the `update()` method. Once the update is complete, you can swap out the old cache with the new cache using the `swapCache()` method.

However, before you begin this update process, check to make sure that the application cache is ready for updating. To do so, check the `status` property of the `applicationCache` object. This property returns one of the values shown in Table 11-1.

Table 11-1: `applicationCache.status` **Values**

Constant	Number	Description
`window.applicationCache.UNCACHED`	0	No cache is available.
`window.applicationCache.IDLE`	1	The local cache is up to date.
`window.applicationCache.CHECKING`	2	The manifest file is being checked for changes.
`window.applicationCache.DOWNLOADING`	3	Safari is downloading changed files and has added them to the cache.
`window.applicationCache.UPDATEREADY`	4	The new cache is ready for updates and to override your existing cache.
`window.applicationCache.OBSOLETE`	5	The cache is obsolete and is being deleted.

For example:

```javascript
if (window.applicationCache.status == window.applicationCache.UNCACHED)
{
    alert("Houston, we have a cache problem");
}
else if (window.applicationCache.status == window.applicationCache.IDLE)
{
    alert("No changes are necessary.");
}
else
{
    alert("Let's do something with the cache.");
}
```

You can assign event handlers to the `applicationCache` object based on the results of the update process. For example:

```javascript
var localCache = window.applicationCache;
localCache.addEventListener("updateready", cacheUpdateReadyHandler, false);
localCache.addEventListener("error", cacheErrorHandler, false);
```

Following are the `applicationCache` events that are supported:

- ❏ `checking`
- ❏ `error`
- ❏ `noupdate`
- ❏ `downloading`
- ❏ `updateready`
- ❏ `cached`
- ❏ `obsolete`

Because I am focused on programmatically performing an update, I will listen for the `updateready` and `error` events:

```javascript
// Handler when local cache is ready for updates
function cacheUpdateReadyHandler()
{
}

// Handler for cache update errors
function cacheErrorHandler()
{
    alert("Houston, we have a cache problem");
}
```

Once these event handlers are defined, you are ready to start the update and swap process.

```javascript
// Handler when local cache is ready for updates
function cacheUpdateReadyHandler()
```

```
{
    localeCache.update();
    localeCache.swapUpdate();
}
```

The `update()` method updates the cache, and `swapUpdate()` replaces the old cache with the new cache you just successfully downloaded.

Checking the Connection Status

When you use application cache, you may have some online processing that you want to disable if you are in offline mode. You can check the connection status in your Web app by checking the `navigator` object's `onLine` property:

```
if (navigator.onLine)
    alert("Online. All services available.")
else
    alert("Offline. Disabling currency rate updates.");
```

Putting It All Together

The following examples demonstrate the application cache in action. Listing 11-1 shows the `index. html` file, and Listing 11-2 provides a listing of the `cacheme.manifest` file. The `localeCache` variable is assigned to the `window.applicationCache`. Handlers are assigned to `applicationCache` events, which determine whether to show a progress indicator of the cache updating process. If the `updateready` event is triggered, the `swapUpdate()` method is called to update to cache.

Listing 11-1: index.html

```
<!DOCTYPE html PUBLIC "-//W3C//DTD XHTML 1.0 Strict//EN"
"http://www.w3.org/TR/xhtml1/DTD/xhtml1-strict.dtd">

<html manifest="cacheme.manifest" xmlns="http://www.w3.org/1999/xhtml">

<head>
    <meta http-equiv="Content-Type" content="text/html; charset=utf-8"
/>
    <meta name="viewport" content="width=device-width,
minimum-scale=1.0, maximum-scale=1.0" />
    <title>Cache Me If You Can</title>

    <script type="text/javascript" language="JavaScript" >

    // Assign var to applicationCache
    var localCache = window.applicationCache;

    // Show progress indicator
    localeCache.addEventListener("progress", cacheProgressHandler,
false);
```

Continued

Listing 11-1: index.html *(continued)*

```javascript
    // Swap cache
    localeCache.addEventListener("updateready",
cacheUpdateReadyHandler, false);
    // Hide progress indicator
    localeCache.addEventListener("cached", cacheNoUpdateHandler,
false);
    localeCache.addEventListener("noupdate", cacheNoUpdateHandler,
false);
    // Show error msg
    localeCache.addEventListener("error", cacheErrorHandler, false);

    // Called on load
    function init()
    {
      if (navigator.onLine)
          document.getElementById("onlineIndicator").textContent =
"online";
        else
          document.getElementById("onlineIndicator").textContent =
"offline";

        localeCache.update();

    }


    // Called when no updates are needed
    function cacheNoUpdateHandler()
    {
        document.getElementById("progressSpinner").style.display =
"none";
    }

    // Called when a cache update is in progress
    function cacheProgressHandler()
    {
      document.getElementById("progressSpinner").style.display =
"table";
    }

    // Called when cache is ready to be updated
    function cacheUpdateReadyHandler()
    {
      localCache.swapCache();
      document.getElementById("progressSpinner").style.display =
"none";
    }

    // Error handler
    function cacheErrorHandler()
    {
        document.getElementById("progressSpinner").style.display =
"none";
```

```
        alert("A problem occurred trying to load the cache.");
    }


    </script>

</head>
<body onload="init()">

    <div id="progressSpinner" style="display:none;">
            <p>Loading...<img src="spinner.gif" alt="Loading..."
width="16" height="16" /></p>
    </div>

    <div id="content">
        <div id="onlineIndicator">online|offline</div>
        <p>This is a test of the Safari on iPhone applicationCache.</p>
        <img src="boy.png" />
    </div>

</body>
</html>
```

Listing 11-2: cacheme.manifest

```
CACHE MANIFEST

#   images
boy.png
spinner.gif
```

Figures 11-1 and 11-2 show this example run in both online and airport modes.

Figure 11-1: Running in online mode

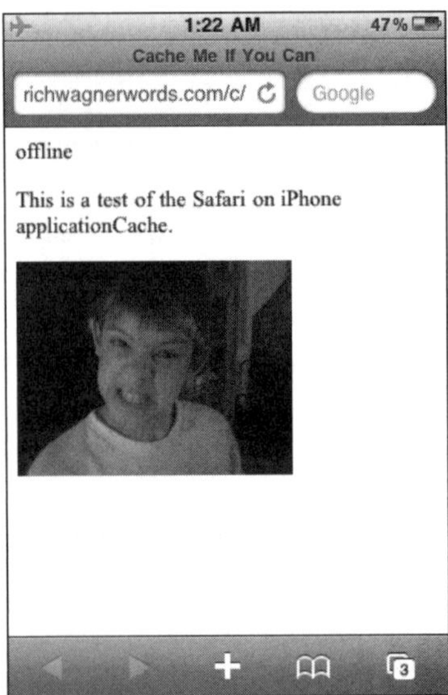

Figure 11-2: Running in offline mode
still works and displays manifest resources

Using Key-Value Storage

In addition to local cache, Safari on iPhone supports HTML 5 key-value storage as a way to provide persistent storage on the client either permanently or within a given browser session. Key-value storage bears several obvious similarities to cookies in client-side storage. However, although cookies are sent back to the server, saved key-value data is not unless you explicitly do so through JavaScript. You also have greater control over the data persistence and window access to that data using key-value storage.

You can specify whether the key values you are saving should be long- or short-term by working with two different JavaScript objects: `localStorage` and `sessionStorage`.

- ❑ Use `localStorage` when you want to store a key-value pair permanently across browser sessions and windows.
- ❑ Use `sessionStorage` to store temporary data within a given browser window and session.

Saving a Key-Value

You can save a key value in one of two ways. First, you can call the `setItem()` method on the `localStorage` or `sessionObject` object.

```
localStorage.setItem(keyName, keyValue);
```

For example, to save a user-inputted value as the `firstName` key for long-term storage, use this:

```
localStorage.setItem("firstName", document.getElementById("first_name").value);
```

Second, a shortcut for saving a value is to treat the key-value name as an actual property of the `local-Storage` or `sessionStorage` objects. For example, the previous example can also be written as follows:

```
localStorage.firstName = document.getElementById("first_name").value);
```

However, if you use this shortcut method, you need to make sure that the name of your key-value is a valid JavaScript token.

Any local database is going to have a maximum capacity, so it is good practice to trap for a possible exception in case the capacity has been reached. To do so, trap for `QUOTA_EXCEEDED_ERR`. For example:

```
try
{
    localStorage.firstName = document.getElementById("first_name").value);
}
catch (error)
{
  if (e == QUOTA_EXCEEDED_ERR)
      alert("Unable to save first name to the database.");
}
```

Whenever you interact with the local storage database, a `storage` event is dispatched from the `body` object. Therefore, to handle this event, you can attach a listener to the `body`:

```
document.body.addEventListener("storage", storageHandler, false);
```

The event object that is passed to the handler enables you to get at various pieces of the transaction (see Table 11-2). For example, the following handler outputs the storage event details to an alert box:

```
function storageHandler(event)
{
    var info = "A storage event occurred [" +
               "url=" + event.url + ", " +
               "key=" + event.key + ", " +
               "new value=" + event.newValue + ", " +
               "old value=" + event.oldValue + ", " +
               "window=" + event.window + "]";

    alert(info);
}
```

Table 11-2: `storage` `event` **Properties**

Property	Description
url	URL of the page that calls the storage object. Returns `null` if the requesting page is not in the same window. Returns `undefined` if calling from a local page.
key	Specifies the key that has been added, changed, or removed.
newValue	Provides the new value for the key.

Continued

Table 11-2: `storage` event **Properties** *(continued)*

Property	Description
`oldValue`	Provides the old value for the key. If no key was previously defined, `null` is returned.
`window`	Reference to the `window` object that called the storage object. Returns `null` if the requesting page is not in the same window.

Loading Key-Value Data

Data can be loaded from the `localStorage` and `sessionStorage` objects by calling the `getItem()` method:

```
var keyValue = sessionStorage.getItem(keyName);
```

For example:

```
var accessCode = sessionStorage.getItem("accessCode");
```

Or, as you would expect, you can access a key value by calling it as a direct property of the respective object:

```
var accessCode = sessionStorage.accessCode;
```

If the key-value that you request is not located, a `null` value is returned:

```
if (sessionStorage.accessCode != null)
{
    var valid = processCode(sessionStorage.accessCode);
}
```

Deleting Key-Value Data

You can remove a specific key-value pair or clear all keys from the local key-value database.

To remove a specific key-value pair, use `removeItem()`:

```
localStorage.removeItem(keyName);
```

To remove all keys, use one or both of the following:

```
// Remove all permanent keys
localStorage.clear()
// Remove all session keys
sessionStorage.clear()
```

Putting It All Together

The following example shows how you can use HTML 5's key-value storage mechanisms to save permanent and temporary values. An input element value is saved as a permanent key-value pair, whereas the selection in a `select` list is saved for the current session only. I want to save the values of these elements each time they change, so I will assign `onchange` handlers to both of these elements:

```
<div>

<p>Define a key value in which you want to save permanently:</p>

<input id='localValue' onchange="setLocalKeyValue('localValue')"/>

<p>Define a key value in which you want to save for this session only:<p>

<select id="sessionValue" onchange="setSessionKeyValue('sessionValue')">
  <option value="UN">(Select State)</option>
  <option value="MA">Massachusetts</option>
  <option value="ME">Maine</option>
  <option value="RI">Rhode Island</option>
  <option value="VT">Vermont</option>
</select>
</div>
```

I also want to add a button to clear the key-values on demand, as well as a status box that acts as a console to output the storage events that are taking place:

```
<button onclick="clearAll()">Clear All</button><br />
<div id="statusDiv">Start session</div>
```

Figure 11-3 shows the page in Safari.

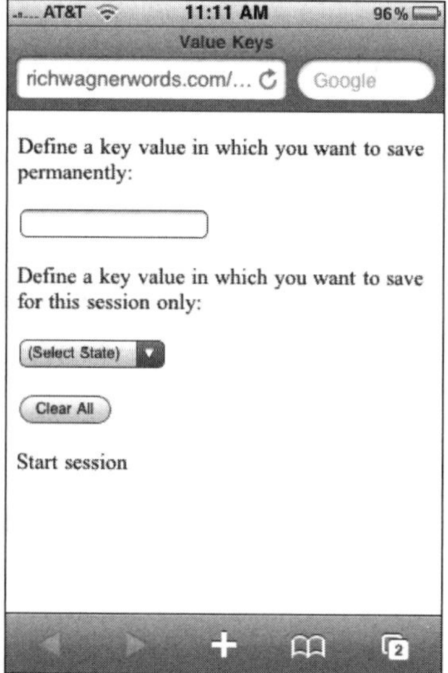

Figure 11-3: Page ready to save data

With the HTML markup ready, I can look at the JavaScript code needed to power this example. To begin, I want to confirm that key-value storage is available, so I do a test on the `localStorage` and `sessionStorage` objects:

```
var localStorageAvail = typeof(localStorage) != "undefined";
var sessionStorageAvail = typeof(sessionStorage) != "undefined";
```

I can then check one or both of these values prior to attempting to save or load storage data.

To save the value of the input element permanently, I add the following function:

```
function setLocalKeyValue(value)
{
    if (localStorageAvail)
    {
        localStorage.setItem(value, document.getElementById(value).value);
        setStatus(document.getElementById(value).id + " saved as a local key.");
    }
}
```

To save the `selectedIndex` of the `select` element as a session-only key-value pair, use the following function:

```
function setSessionKeyValue(value)
{
    if (sessionStorageAvail)
    {
        sessionStorage.setItem(value, document.getElementById(value)
        .selectedIndex);
        setStatus(document.getElementById(value).id + " saved as a session key.");
    }
}
```

To retrieve the values of these key-value pairs, I add a loading function:

```
function loadValues()
{
    if (localStorage.localValue)
        document.getElementById('localValue').value = localStorage.localValue;
    if (sessionStorage.sessionValue)
        document.getElementById('sessionValue').selectedIndex =
        sessionStorage.sessionValue;
}
```

Next, to clear all the local values, a `clearStorage()` function is defined to remove key-value pairs from both `sessionStorage` and `localStorage` objects as well as the UI fields:

```
function clearStorage()
{
    // Clear local database
    sessionStorage.clear();
```

```
    localStorage.clear();

    // Clear UI as well
    document.getElementById('localValue').value = "";
    document.getElementById('sessionValue').selectedIndex = 0;
}
```

Listing 11-3 shows the full source code for the HTML file, which includes additional functions and event handlers to tie everything together.

Listing 11-3: index.html

```
<!DOCTYPE html PUBLIC "-//W3C//DTD XHTML 1.0 Strict//EN"
"http://www.w3.org/TR/xhtml1/DTD/xhtml1-strict.dtd">
<html xmlns="http://www.w3.org/1999/xhtml">
<head>
    <meta http-equiv="Content-Type" content="text/html;
charset=utf-8" />
    <meta name="viewport" content="width=device-width,
minimum-scale=1.0, maximum-scale=1.0" />
    <title>Value Keys</title>

    <script language="JavaScript" type="text/javascript">

        // Check to see whether or not key-value storage is
        available
        var localStorageAvail = typeof(localStorage) !=
        "undefined";
        var sessionStorageAvail = typeof(sessionStorage) !=
        "undefined";

        // Assign event listeners
        window.addEventListener("onload", init, false);
        window.addEventListener("onbeforeunload",
        beforeUnloadHandler, false);

        // Called on load
        function init()
        {
            // If no storage is available
            if (!localStorageAvail || !sessionStorageAvail)
              setStatus("Key value storage is not supported.");
            // Otherwise, assign handler and retrieve any
            // previously stored values
            else
            {
                document.body.addEventListener("storage",
                storageHandler, false);
                loadValues();
            }
        }

        // Save local key
```

Continued

Listing 11-3: index.html *(continued)*

```javascript
// value = id of the element whose value is being saved
function setLocalKeyValue(value)
{
    if (localStorage)
    {
        localStorage.setItem(value,
        document.getElementById(value).value);
        setStatus(document.getElementById(value).id +
        " saved as a local key.");
    }
}


// Save session key
// value = id of the element whose value is being saved
function setSessionKeyValue(value)
{
    if (sessionStorage)
    {
        sessionStorage.setItem(value,
        document.getElementById(value).selectedIndex);
        setStatus(document.getElementById(value).id +
        " saved as a session key.");
    }
}



// Loads key-value pairs from local storage
function loadValues()
{
    // Is localValue defined? If so, then assign its
    // value to the text box.
    if (localStorage.localValue)
        document.getElementById('localValue').value =
        localStorage.localValue;
    // Is sessionValue defined? If so, then assign its
    // value as the index to the sessionValue select.
    if (sessionStorage.sessionValue)
        document.getElementById('sessionValue')
        .selectedIndex = sessionStorage.sessionValue;
}

// Clear all key-value pairs
function clearStorage()
{
    // Clear local database
    sessionStorage.clear();
    localStorage.clear();

    // Clear UI as well
    document.getElementById('localValue').value = "";
    document.getElementById('sessionValue').
    selectedIndex = 0;
```

```javascript
            }

            // Save current state
            function saveChanges()
            {
                setLocalValue('localValue');
                setSessionKeyValue("sessionValue");
                // Used to return to the window
                return null;
            }

            // Be sure to save changes before closing a window
            function beforeUnloadHandler()
            {
                return saveChanges();
            }

            // Listener for all storage events. Outputs to the status div.
            function storageHandler(event)
            {
                var info = "A storage event occurred [" +
                        "url=" + event.url + ", " +
                        "key=" + event.key + ", " +
                        "new value=" + event.newValue + ", " +
                        "old value=" + event.oldValue + ", " +
                        "window=" + event.window + "]";

                setStatus(info);
            }

            //  Utility function that outputs specified text to the status
            // div
            function setStatus(statusText)
            {
                var para = document.createElement("p");
                para.appendChild(document.createTextNode(statusText));
                document.getElementById("statusDiv").appendChild(para);
            }

        </script>
        </head>

        <body onload="init()">
        <div>
        <p>Define a key value in which you want to save permanently:</p>

        <input id='localValue' onchange="setLocalKeyValue('localValue')"/>

        <p>Define a key value in which you want to save for this session
        only:</p>

        <select id="sessionValue" onchange="setSessionKeyValue
        ('sessionValue')">
          <option value="UN">(Select State)</option>
```

Continued

Listing 11-3: index.html *(continued)*

```html
            <option value="MA">Massachusetts</option>
            <option value="ME">Maine</option>
            <option value="RI">Rhode Island</option>
            <option value="VT">Vermont</option>
        </select>
    </div>

    <p>
    <button onclick='clearAll()'>Clear All</button><br />
    </p>
    <div id="statusDiv">Start session</div>
    </body>
    </html>
```

The following figures demonstrate this example when run. Figure 11-4 shows the data being entered on-screen. Figure 11-5 is what's shown after pressing the Refresh button within the current session. As you can see, both values are retained. However, after closing out that window and calling the URL again, the session value is cleared, as shown in Figure 11-6.

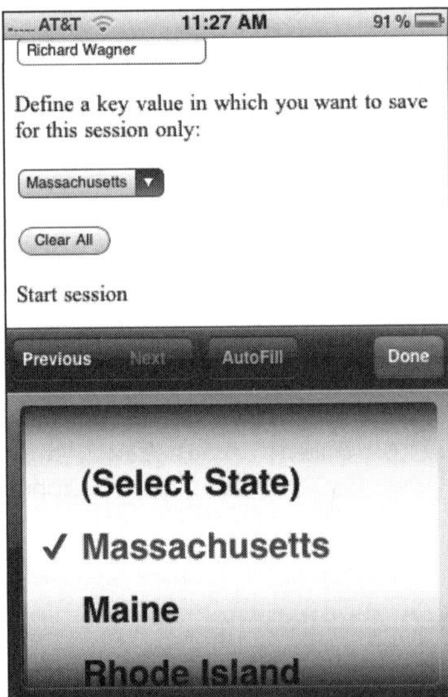

Figure 11-4: Values being stored as key-value pairs

When you are working with key-value storage, I recommend using the Web Inspector that is provided with the Windows and Mac versions of Safari (accessed from the Develop menu). The Databases panel (see Figure 11-7) enables you to see the current state of the key-value pairs for the page you are working with.

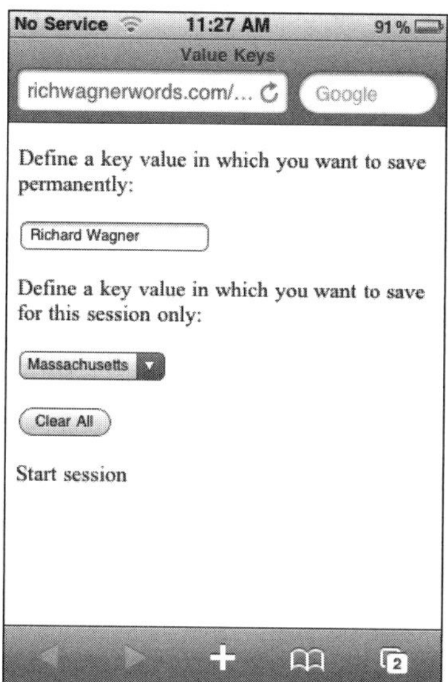

Figure 11-5: Refreshing the page in the current session retains both permanent and temporary values.

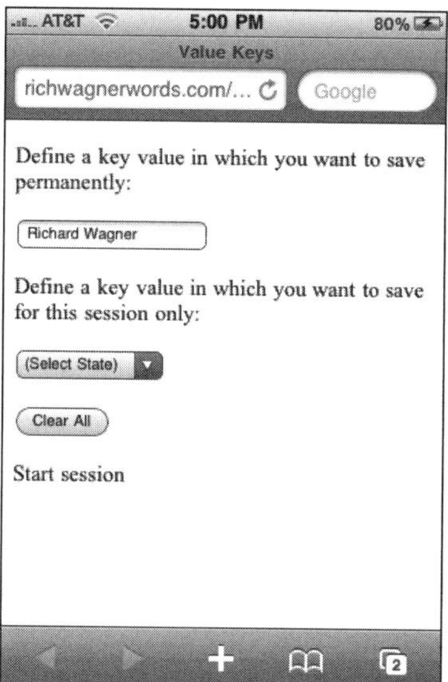

Figure 11-6: A new browser session shows just the permanent key-value.

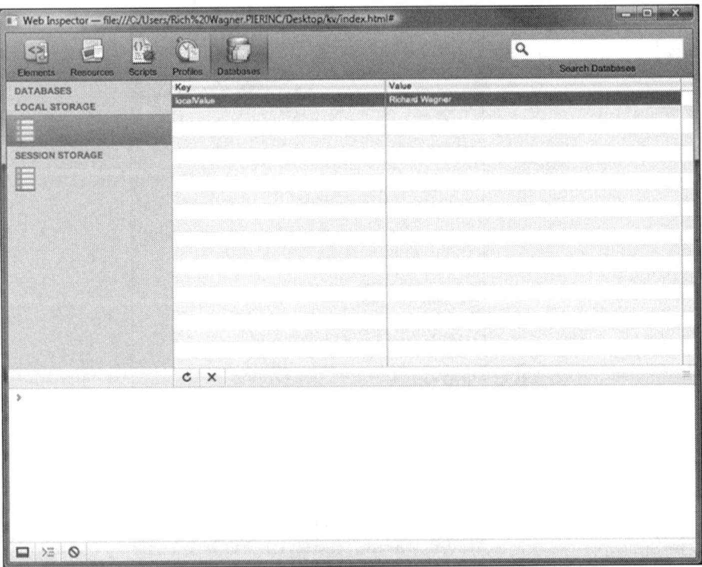

Figure 11-7: Interact with key-value pairs in the Databases panel of the Safari Web Inspector

Going SQL with the JavaScript Database

Once you begin to develop more substantial Web applications, you may easily have local storage needs that go beyond simple caching and key-value pair persistence. Perhaps your app stores application data locally and periodically synchs with a database on a backend server. Or maybe your Web app uses a local database as its sole data repository. Using HTML 5's database capabilities, you can access a local SQLite relational database right from JavaScript to create tables, add and remove records, and perform queries.

The SQLite database is an SQL relational database. For full details on how to work with SQL to create, edit, and query your data, see Robert Vieira's *Beginning Microsoft SQL Server 2008 Programming* (Wrox, 978-0-470-25701-2).

Opening a Database

Your first step is to open a database by calling the `openDatabase()` method of the `window` object:

```
var db = window.openDatabase(dbName, versionNum, displayName, maxSize);
```

The `dbName` parameter is the database name stored locally, `versionNum` is its version number, `displayName` is the display name of the database used by the browser (if needed), and `maxSize` is the maximum size in bytes that the database will be. (Additional size requires user confirmation.) For example:

```
var db = window.openDatabase("customers", "1.0", "Customer database", 65536);
```

Once the database is opened, you can perform various tasks on it.

Each task you perform must be part of a transaction. A database transaction can include one or multiple SQL statements and is set up as follows:

```
db.transaction(transactionFunction, errorCallbackFunction,
successCallbackFunction);
```

Querying a Table

Suppose you wanted to perform a simple query on a customer table in your database. You could set up the query inside of a transaction, such as what is shown here:

```
var db = window.openDatabase("customers", "1.0", "Customer database", 65536);

if (db != null)
{
    var updateSqlStr = "SELECT * FROM customers";
    db.transaction( function(transaction) { transaction.executeSql
(updateSqlStr) },  successHandler, errorHandler);
}


function successHandler(result)
{
    if (result.rows.length > 0)
    {
        for (var i = 0; i < result.rows.length; i++)
        {
            var r = results.rows.item(i);
            alert(r["first_name"] + " was retrieved from the database.");
        }
    }
}

function errorHandler(error)
{
    alert("An error occurred when trying to perform a query.");
}
```

In this example, the `db.transaction` calls an SQL execute statement to return all the customers from the database. The `successHandler()` function is called when the database is successful. The `errorHandler()` function is called if the operation fails.

The result object returned contains a rows array that contains each record in the returning set. You can access each of the fields by specifying its field name inside the brackets.

Summary

Safari on iPhone provides full support for HTML 5's offline capabilities. As a result, you can now create Web apps that can work even when the user does not have direct access to the Internet. In this chapter, I showed you how to work with Safari's offline capabilities. I began by explaining the offline cache and how to configure one through a manifest file for your Web app. The chapter continued with a discussion of key-value storage, enabling you to store both permanent and temporary user data in a way that goes far beyond traditional cookie storage. Finally, I introduced you to how you can access an SQLite relational database from within JavaScript.

12

Enabling and Optimizing Web Sites for the iPhone and iPod Touch

Oh, the irony. On the same day that I began writing a chapter on enabling Web sites for iPhone and iPod touch, I realized firsthand the frustration of browsing sites that just don't work with my iPhone. My boys and I were watching the third quarter of a Monday Night Football game when the electricity suddenly went out because of a town-wide outage. Because my son's favorite team was playing, he was frantic. *What's happening in the game? Are the Titans still winning?* I immediately pulled out my iPhone and confidently launched Safari in search of answers. But upon going to NFL.com, I discovered that its live updating scoreboard is Adobe Flash media. I was left with a gray box with a Lego-like block in its place. I then pointed the browser to the official Tennessee Titans site, only to discover useless Lego blocks scattered across its front page as well. We then spent the rest of the outage scouring the Web, looking for a sports site to help us.

If you manage a Web site, the iPhone introduces a whole new way of thinking in the design and development of a site. In the past, you could design a minimalist, text-only style sheet for mobile users — fully expecting your normal Web site to be viewed only by desktop browsers. However, expectations of iPhone and iPod touch users are not so modest. They are expecting to view the *full Web* in the palm of their hands. Therefore, as you design and develop your Web site, you will want to consider the level of support you want to provide for these Apple devices — whether to offer mere compatibility, device friendliness, or even a design specifically targeting them. This chapter goes over the four tiers of enabling your Web site for Safari on iPhone:

- ❏ Tier 1: Compatibility
- ❏ Tier 2: Navigation friendliness
- ❏ Tier 3: Device-specific style sheets
- ❏ Tier 4: Dedicated alternative site

Tier 1: iPhone/iPod touch Compatibility

The first tier of support for iPhone is simply making your Web site work inside Safari on iPhone. Fortunately, because Safari is a sophisticated browser, far closer in capability to a desktop than a typical mobile browser, this is usually not problematic. However, there are some gotchas that you'll encounter. These include

❑ Adobe Flash media, Java applets, and plug-ins are not supported.

❑ You cannot use the CSS property `position:fixed`.

❑ JavaScript functions `showModalDialog()` and `print()` do not function under Safari on iPhone.

❑ Downloads and uploads (including HTML element `input type="file"`) are not supported.

Given its widespread popularity and desktop install base, Flash is the thorniest incompatibility for many Web designers and developers. Until the iPhone's release, Flash support was typically considered a given except for a relatively small percentage of users. In fact, many designers could take it for granted that if a user was coming to their site without Flash support, they probably were not a target visitor anyway, so they could either ignore them or refer them to the Adobe download page. However, with the release of iPhone and iPod touch, those assumptions are now invalid. Web designers are forced to rethink their site's reliance on a technology that they had become dependent upon. Figures 12-1 and 12-2 demonstrate the harsh reality, in which a state-of-the-art Web site that looks amazing in Safari for Mac OS X never accounts for iPhone users.

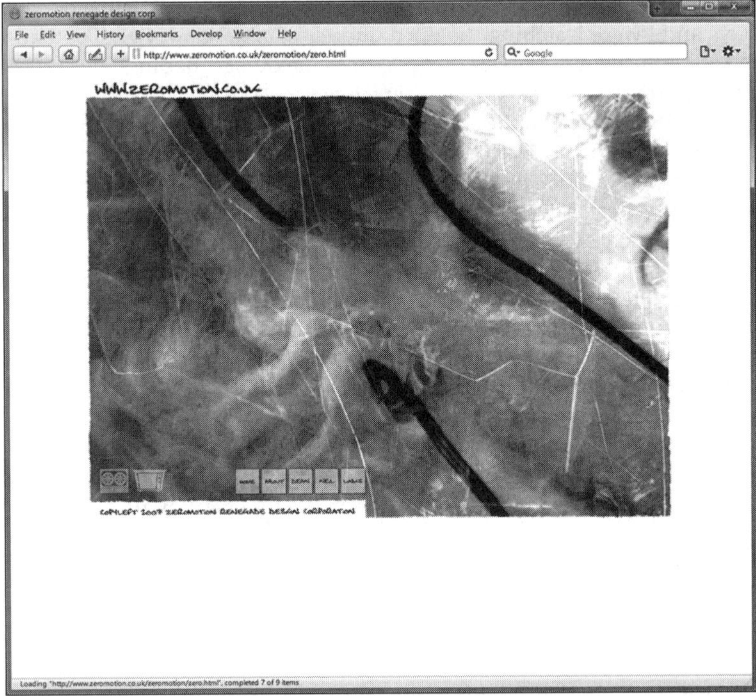

Figure 12-1: Flash-based site that attracts desktop users...

Figure 12-2:...leaves iPhone users out in the cold.

Therefore, if you plan to use Flash for an interactive portion of a page, you should plan to degrade gracefully to a static graphic or alternative content. At a minimum, you should place a disclaimer over Flash content. It's not ideal, but it is better than the Lego block. Or, if you have a Flash-driven site (such as the one shown in Figure 12-1), you should consider an alternative HTML site or, if warranted, even an iPhone-specific site.

To detect Flash support, one solution is to use SWFObject, an open source JavaScript library that is used for detecting and embedding Flash content (available at `blog.deconcept.com/swfobject`). SWFObject is not iPhone specific, but encapsulates the Flash Player detection logic, making it easy for you to degrade gracefully for Safari on iPhone. For example, the following code will display a Flash file for Flash-enabled desktop browsers but will display a splash PNG graphic for non-Flash visitors, including iPhone and iPod touch users:

```
<!DOCTYPE html PUBLIC "-//W3C//DTD XHTML 1.0 Strict//EN"
           "http://www.w3.org/TR/xhtml1/DTD/xhtml1-strict.dtd">
<html xmlns="http://www.w3.org/1999/xhtml">
<head>
<title>Company XY Home Page</title>
<meta name="viewport" content="width=780">
<script type="text/javascript" src="swfobject.js"></script>
</head>
<body>
<div id="splashintro">
  <a href="more.html"><img src="splash_noflash.png"/></a>
</div>
<script type="text/javascript">
   var so = new SWFObject("csplash.swf", "company_intro", "300", "240", "8",
   "#338899");
```

```
        so.write("splashintro");
    </script>
    </body>
    </html>
```

As you can see, the `swfobject.js` library file is added to the home page. When Flash is available, the script replaces the content of the `splashintro` div with Flash media. When Flash is not supported, appropriate content is substituted inside the `splashintro` div.

Therefore, at a minimum, you should seek to make your Web site fully aware and compatible for Safari on iPhone users.

Tier 2: Navigation-Friendly Web Sites

Once your Web site degrades gracefully for iPhone users, you have achieved a base level of support for Apple mobile devices. However, although Safari on iPhone users may be able to see all the content on a Web site, they might still have trouble navigating and reading it. A wide section of text, for example, becomes a stumbling block for iPhone and iPod touch users to read because horizontal scrolling is required when users zoom in to read it. With this in mind, the second tier of support is to structure the site in a manner that is easy for Safari to zoom and navigate.

Working with the Viewport

A *viewport* is a rectangular area of screen space within which a Web page is displayed. It determines how content is displayed and scaled to fit onto the iPhone. Using the viewport is analogous to looking at a panoramic scenic view of a mountain range through a camera zoom lens. If you want to see the entire mountainside, you zoom out using the wide angle zoom. As you do so, you see everything, but the particulars of each mountain become smaller and harder to discern. Or, if you want to see a close-up picture of one of the peaks, you zoom in with the telephoto lens. Inside of the camera's viewfinder, you can no longer see the range as a whole, but the individual mountain is shown in terrific detail. The `viewport` meta tag in Safari works much the same way, allowing you to determine how much of the page to display, its zoom factor, and whether you want users to zoom in and out or whether they need to browse using one scale factor.

The way in which Safari renders the page is based largely on the `width` (or `initial-scale`) property of the `viewport` meta tag. With no `viewport` tag present, Safari considers the Web page it is loading as 980 pixels in width, and then it shrinks the page scaling so that the entire page width can fit inside the 320-pixel viewport (see Figure 12-3). Here is the default declaration:

```
<meta name="viewport" content="width=980;user-scalable=1;"/>
```

Suppose your Web site is only 880 pixels wide. If you let Safari stick with its default 980-pixel setting, the page is scaling more than it needs to. Therefore, to adjust the viewport magnification, you can specify a width optimized for your site:

```
<meta name="viewport" content="width=880"/>
```

Figures 12-4 and 12-5 show the noticeable difference between a 980- and an 880-width viewport for an 880-pixel width site.

Figure 12-3: A 980px–wide Web page scaled to fit in iPhone.

With this declaration, instead of trying to fit 980 pixels into the 320 pixels of width, the site only needs to shrink 880 pixels. Less scaling of content is needed (.363 scale instead of .326), making the site easier to use for iPhone and iPod touch users. Note that the `viewport` meta tag will not affect the rendering of the page in a normal desktop browser.

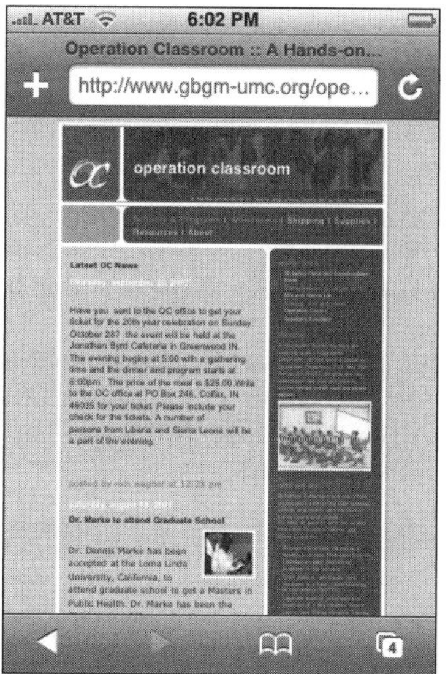

Figure 12-4: The default width creates empty space on the edges.

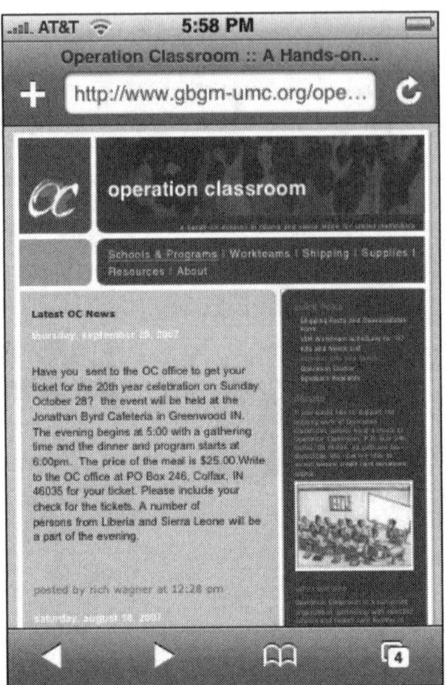

Figure 12-5: The viewport is adjusted to better fit the Web page.

In addition to the `width` property, you can programmatically control the scale of the viewport when the page initially loads through the `initial-scale` parameter. For example, if you wanted to set the initial scale to be .90, the declaration would be

```
<meta name="viewport" content="initial-scale=.9;user-scalable=1;"/>
```

Once the page loads, however, users can change the scale factor as they want using pinch and double-tap gestures as long as the `user-scalable` property is set to `true` (the default). If you want to limit the scale range, you can use the `minimum-scale` and `maximum-scale` properties:

```
<meta name="viewport" content="initial-scale=.9;maximum-scale=1.0;
minimum-scale=.8;user-scalable=1;"/>
```

In this way, users can pinch and zoom, but only to the extent that you want to allow.

If you develop a site or app specifically for iPhone, you'll want to size the page to the viewport by setting the `width=device-width` (`device-width` is a constant) and `initial-scale=1.0`. Because the scale is 1.0, you don't want users to be able to rescale the application interface, so the `user-scalable` property should be disabled. Here's the declaration:

```
<meta name="viewport" content="width=device-width; initial-scale=1.0;
maximum-scale=1.0; user-scalable=0;">
```

Table 12-1 lists the `viewport` properties. You don't need to set every property. Safari infers values based on the properties you have set.

Keep in mind that the `width` attribute does not refer to the size of the Safari browser window, but instead the perceived size of the page in which Safari shrinks down to be displayed properly on the mobile device.

Table 12-1: viewport Meta Tag Properties

Property	Default Value	Minimum Value	Maximum Value	Description
`width`	980	200	10000	Width of viewport
`height`	Based on aspect ratio	223	10000	Height of viewport
`initial-scale`	Fit to screen	Minimum scale	Maximum scale	Scale to render when page loads
`user-scalable`	1 (yes)	0 (no)	1 (yes)	If yes, user can change scale through pinch and double-tap
`minimum-scale`	0.25	>0	10	Use to set the lower end for scaling
`maximum-scale`	1.6	>0	10	Use to set the higher end for scaling

Although it's not generally recommended, you can specify the width of the content to be greater than the viewport width, but that requires the users to scroll horizontally.

Note that current versions of the iPhone OS (1.1.1 and above) support two width and height constants: `device-width` (width of device in pixels, or `320`) and `device-height` (height of device in pixels, or `480`).

Turning Your Page into Blocks

One of the most important ways to make your Web site friendly for iPhone users is to turn your Web page into a series of columns and blocks. Columns make your page readable like a newspaper and help you avoid wide blocks of text that cause users to horizontally scroll left and right to read.

When an element is double-tapped, iPhone finds its closest block (`div`, `ol`, `ul`, `table`, and so on) or image ancestor. If a block is found, Safari zooms the content to fit the block's content based on the `viewport` tag's `width` property value and then centers it. If an image is tapped, Safari zooms to fit the image and centers it. If an image is already zoomed, zoom out occurs.

Figure 12-6 shows a sample page with a relatively simple structure, but one that makes it difficult for the iPhone to zoom in on it. The table is defined at a fixed width of 1000px, and the first column takes up 875px of that space. The text above the table spans the full document width, but because it is outside of a block, Safari can do no zooming when the text is double-tapped. The user is forced to go to landscape mode and pinch to get readable text, but it still scrolls off the right of the screen (see Figure 12-7).

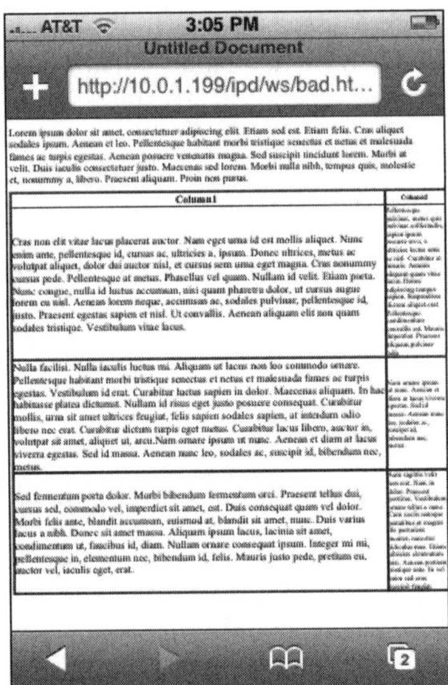

Figure 12-6: Unfriendly page on page load

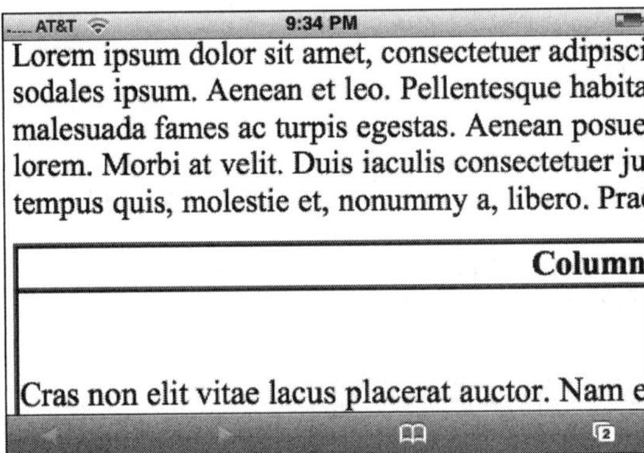

Figure 12-7: Zooming to a cell

However, with a few simple tweaks, you can transform the page into something far easier for iPhone and iPod touch to work with. First, you can add a `viewport` meta tag to gain greater control over the width:

```
<meta name="viewport" content="width=950"/>
```

Second, you can enclose the paragraph into a `div` block element and transform it into a column (say 50 percent of the page):

```
<div style="width:50%">
</div>
```

In the real world, you would obviously want to tailor the entire page design around a more column-based approach.

Third, you can size by percentage (90 percent of width) rather than the fixed width of 1000px:

```
<table width="90%" border="1" cellspacing="1" cellpadding="1">
  <tr>
    <th width="75%" valign="top" scope="col"><div align="center">Column1</div></th>
    <th width="25%" valign="top" scope="col">Column2</th>
  </tr>
  <tr>
  .
</table>
```

Even with these rudimentary changes, the page becomes easier to browse when you double-tap the page, as shown in Figure 12-8.

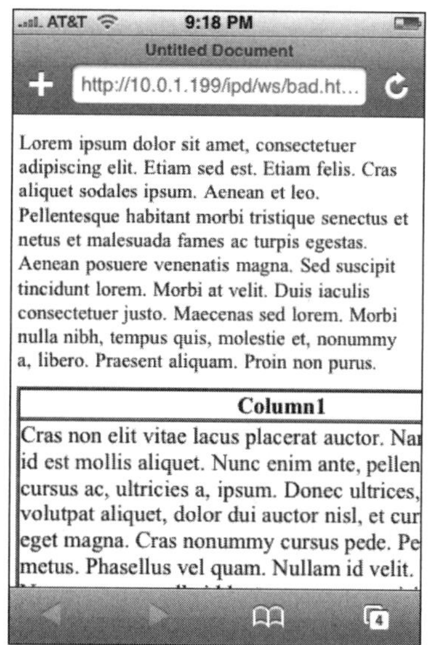

Figure 12-8: The text block is now readable

Figure 12-9 shows the model block-based Web page that is easily navigated with the double-tap and pinch gestures of iPhone.

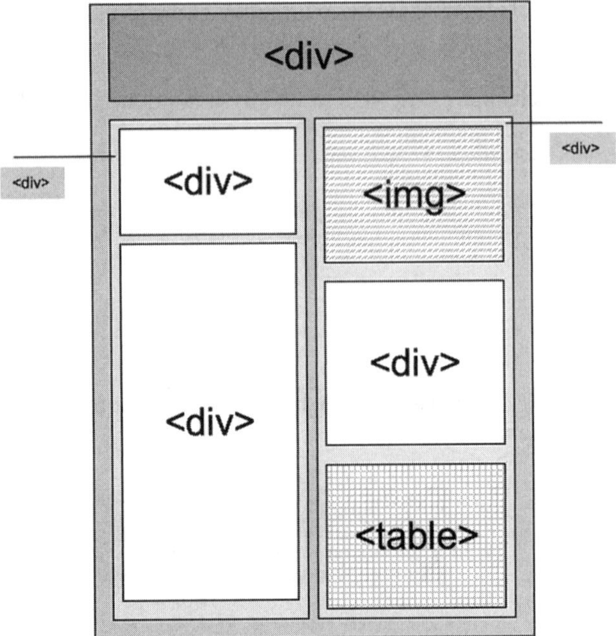

Figure 12-9: The prototype structure of an easy-to-browse page

As you can see, a big part of the strategy being employed is to use percentage-based sizing rather than fixed sizes for page elements.

Defining Multiple Columns (Future Use)

The latest versions of WebKit (Safari) and Mozilla-based browsers provide support for new CSS3 properties that enable you to create newspaper-like, multicolumn layouts. For a content block, you can specify the number of columns, the width of the columns, and the gap between them. Because not all browsers currently support multiple columns, these style properties are prefixed with -webkit and -moz:

```
-webkit-column-count: 2;
-moz-column-count: 2;
-webkit-column-width: 200px;
-moz-column-width: 200px;
-webkit-column-gap: 13px;
-moz-column-gap: 13px;
```

Tier 3: Custom Styling

An iPhone user can navigate a Tier 2 Web site with double-tap, pinch, and flick gestures, but it is not necessarily easy or enjoyable to do so. Panning and scrolling across the screen can quickly become tiresome after the excitement over the "full Web" wears off. Users will find themselves returning to sites that provide a richer, more tailored experience for Safari. The easiest way to do this is to create custom styles specifically for iPhone.

Media Queries

If you want to specify a style sheet for iPhone usage, you can use a CSS3 media query. iPhone does not support the `handheld` or `print` media types, but instead looks for the `screen` media type. You can then use the `link` element to set specific styles for iPhone by looking only for devices that support the screen and have a maximum width of 480px:

```
<link media="only screen and (max-device-width: 480px)"
  rel="stylesheet" type="text/css" href="iphone-ipod.css"/>
```

Or, to set iPhone-specific styles inside a particular CSS style sheet, you could use

```
@media only screen and (max-device-width: 480px)
{
   /* Add styles here */
}
```

The `link` element and the CSS rule would apply only to devices that have a maximum width of 480 pixels. Browsers that do not support the `only` keyword ignore the rule anyway. However, under certain situations, earlier versions of Internet Explorer (versions 6 and 7) fail to ignore this rule and render the page anyway using the iPhone-specific style sheet. You need to guard against this possibility by using IE's conditional comments:

```
<!--[if !IE]>-->
<link media="only screen and (max-device-width: 480px)"
  rel="stylesheet" type="text/css" href="iphone-ipod.css"/>
<!--<![endif]-->
```

Internet Explorer now ignores this link element, because the `[if !IE]` indicates that the enclosed code should only be executed outside of IE.

Therefore, if you want to have a default style sheet for normal browsers and a custom style sheet for iPhone users, use the following combination:

```
<link media="screen and (min-device-width: 481px)"
  rel="stylesheet" type="text/css" href="default.css"/>
<!--[if !IE]>-->
<link media="only screen and (max-device-width: 480px)"
  rel="stylesheet" type="text/css" href="iphone-ipod.css"/>
<!--<![endif]-->
```

Text Size Adjustment

Normally, the font size of a Web page adjusts automatically when the viewport is adjusted. For instance, after a double-tap gesture, Safari looks at the zoomed width of the content block and adjusts the text to zoom in proportion. This behavior makes the text easier to read for typical uses, though it can affect absolute positioning and fixed layouts. However, if you want to prevent the text from resizing, use the following CSS rule:

```
-webkit-text-size-adjust: none;
```

In general, for most Web site viewing, keep this property enabled. For iPhone-specific applications in which you want more control over scaling and sizing, disable this option.

Case Study

Consider a case study example: the Web site of Operation Classroom, a nonprofit organization doing educational work in Africa. Keep in mind that the style sheet of each Web site needs to be optimized in a unique manner, but this case study demonstrates some of the common issues that will crop up.

Figure 12-10 displays a page from the site with a basic `viewport` meta tag set at `width=780`, which gives it the best scale factor for the existing page structure. However, even when the viewport setting is optimized, a user still needs to double-tap to read the text on the page. What's more, the top-level links are difficult to tap unless you pinch and zoom first.

Figure 12-10: The prototype structure of an easy-to-browse page

However, by creating an iPhone and iPod touch–specific style sheet, you can transform the usability site for Mobile Safari users without impacting the HTML code. Looking at the page (see Figure 12-11), you'll notice that several transformations need to occur:

- ❑ Shrink the page width.
- ❑ Shrink the Operation Classroom logo at the top of the page.
- ❑ Increase the font size for the menu links, page header, rabbit trail links, and body text.
- ❑ Move the sidebar so it's below the body text.

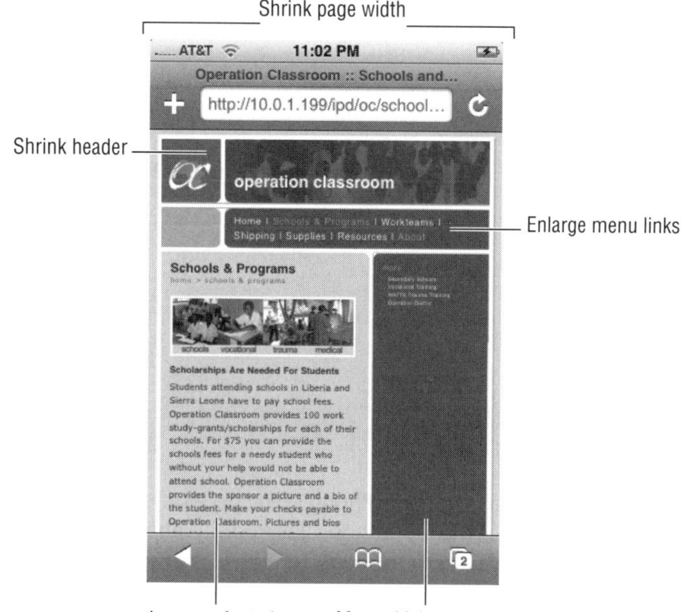

Figure 12-11: Transforming the structure using CSS

As a first step, add a media query to the document head of each page in the site:

```
<link media="screen and (min-device-width: 481px)"
  rel="stylesheet" type="text/css" href="css/oc-normal.css"/>
<!--[if !IE]>-->
<link media="only screen and (max-device-width: 480px)"
  rel="stylesheet" type="text/css" href="css/oc-iphone-ipod.css"/>
<!--<![endif]-->
```

Next, inside of the HTML files, change the `viewport` meta tag to a smaller width:

```
<meta name="viewport" content="width=490"/>
```

The 490px width is wide enough to be compatible with the existing site structure but small enough to minimize the scaling.

That's all the work that you need to do to the HTML files.

To create the new custom style sheet, begin with the default style sheet already being used and then save as a new name — `oc-iphone-ipod.css`. Your first task is to change the width of the document from 744px to 490px. Here's the updated style:

```
@media all {
  #wrap {
    position:relative;
    top:4px;
    left:4px;
```

```
      background:#ab8;
      width:490px;
      margin:0 auto;
      text-align:left;
   }
```

Next, you change the original `font-size:small` property defined in `body` to a more specific pixel size:

```
body {
   background:#cdb;
   margin:0;
   padding:10px 0 14px;
   font-family: Verdana,Sans-serif;
   text-align:center;
   color:#333;
   font-size: 15px;
   }
```

Although this size is not as large as what an iPhone Web app would use, it is the largest font size that works with the current structure of the Operation Classroom Web site. Fortunately, the *rabbit trail* (pathway) and page header fonts are relative to the body font:

```
#pathway {
   margin-top:3px;
   margin-bottom: 25px;
   letter-spacing: .18em;
   color: #666666;
   font-size: .8em;
}
#pageheader {
   font-family:Helvetica,Arial,Verdana,Sans-serif;
   font-weight: bold;
   font-size: 2.2em;
   margin-bottom: 1px;
   margin-top: 3px;
}
```

The next issue is to shrink the size of the banner at the top of the page. Here's the style for the banner text:

```
#banner-text{
   background:url("./images/bg_header.jpg") no-repeat left top;
   margin:0;
   padding:40px 0 0;
   font:bold 275%/97px Helvetica,Arial,Verdana,Sans-serif;
   text-transform:lowercase;
   }
```

The two properties you need to try to shrink are the `padding` and the `font` size. Here's a workable solution:

```
#banner-title {
   background:url("./images/bg_header.jpg") no-repeat left top;
```

```
margin:0;
padding:10px 0 10px;
font: Bold 35px Helvetica,Arial,Verdana,Sans-serif;
text-transform:lowercase;
}
```

The final and perhaps most important change is to enable the sidebar to follow the main text rather than float alongside it. Here's the original definition:

```
#sidebar {
    background:#565 url(".images/corner_sidebar.gif") no-repeat left top;
    width: 254px;
    float: right;
    padding:0;
    color:#cdb;
    }
```

To move the sidebar content below the main body text, remove the `float` property and add a `clear: both` declaration to prevent the sidebar from side wrapping. Also change the small width of 254px to 100 percent, which enables it to take up the entire content of the `content div`. Here's the code:

```
#sidebar {
    background:#565 url("./images/corner_sidebar.gif") no-repeat left top;
    width:100%;
    clear: both;
    padding:0;
    color:#cdb;
}
```

Figures 12-12, 12-13, and 12-14 show the results of the transformation.

Figure 12-12: The top banner is smaller, but the link sizes are larger.

Figure 12-13: Text is easily readable without the need for double-tap or pinch gestures.

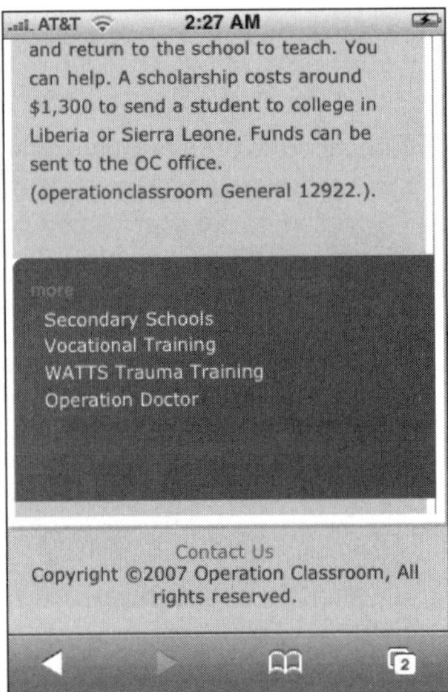

Figure 12-14: Sidebar content now follows main body text.

Tier 4: Parallel Sites

Unless you are creating an iPhone or iPod touch application, developing for Tier 2 or 3 support provides sufficient support for most sites. However, you might find a compelling need to actually develop a site specifically written for iPhone (which I am referring to as a Tier 4 site). The content may be the same, but it needs to be structured in a manner discussed throughout this book.

Avoid Handcuffs, Offer Freedom

If you are going to offer an iPhone version of your site, you want to give your users the freedom to choose between the customized site and your normal site. Don't auto-redirect based on user agent. Because Safari on iPhone can navigate a normal Web site, you should always allow users to make the decision themselves. Therefore, when users access the home page on your iPhone, it clearly notifies them of the alternative site but does not handcuff them into using it.

To add a similar functionality to a Web site, begin by adding an empty `div` element at the top of your content, just below the top menu:

```
<div id="iphone-ipod-notify"></div>
```

This element serves as the placeholder for the message you will display to iPhone users.

Next, add the following script:

```
<script type="application/x-javascript">
function isAppleMobile() {
  result ((navigator.platform.indexOf("iPhone") != -1) ||
          (navigator.userAgent.indexOf('iPod') != -1))
}
function init() {
  if ( isAppleMobile ) {
      var o = document.getElementById( 'iphone-ipod-notify' );
      o.innerHTML = "<h1 style='text-align:center;border: 1px solid #a23e14;
-webkit-border-radius: 10px;'><a href='iphone-ipod-index.html'>Tap here to
go to our<br/>iPhone/iPod touch web site.</a></h1>";
    }
}
</script>
```

The `init()` function calls `isAppleMobile()` to determine whether the user agent is an Apple mobile device. If it is, HTML content is added to the placeholder `div` element. If it's not, nothing is added. The `init()` function is then called from the `onload` handler of the `body`. Figure 12-15 shows the results when viewed from an iPhone.

Figure 12-15: Offering a freedom of choice to your users.

Transform a Site to an iPhone Design

Once you decide to create a companion site specifically for Safari users, you have to decide how exist-ing content best fits inside of an iPhone UI design. You need to determine whether you want to create your own custom design or model after the standard edge-to-edge navigation. The edge-to-edge design scheme works well for many Web sites, as you'll see here.

As a case study, you'll turn once again to the Operation Classroom Web site, the home page of which is shown in Figure 12-16. Several aspects of this site lend themselves to using the edge-to-edge navigation UI. First, the site hierarchy could easily be converted to a series of nested list items. Second, the news entries and quick links entries work great as lists.

Using the iUI framework and the cUI extension (see Chapter 10, "Integrating with iPhone Services"), you'll create a new HTML page containing the top-level menu. Here's the initial code:

```
<body>
    <!--Top iUI toolbar-->
    <div class="toolbar">
        <h1 id="pageTitle"></h1>
        <a id="backButton" class="button" href="#"></a>
    </div>
    <!--Top-level menu-->
    <ul id="home" title="OC for iPhone" selected="true">
        <li><a href="#news">News</a></li>
```

```
        <li><a href="#quick-links">Quick Links</a></li>
        <li><a href="#schools-programs">Schools and Programs</a></li>
        <li><a href="#workteams">Workteams</a></li>
        <li><a href="#shipping">Shipping</a></li>
        <li><a href="#supplies">Supplies</a></li>
        <li><a href="#resources">Resources</a></li>
        <li><a href="#about">About OC</a></li>
        <li><a href="index.html" target="_self">Return to Regular web
        Site</a></li>
    </ul>
</body>
```

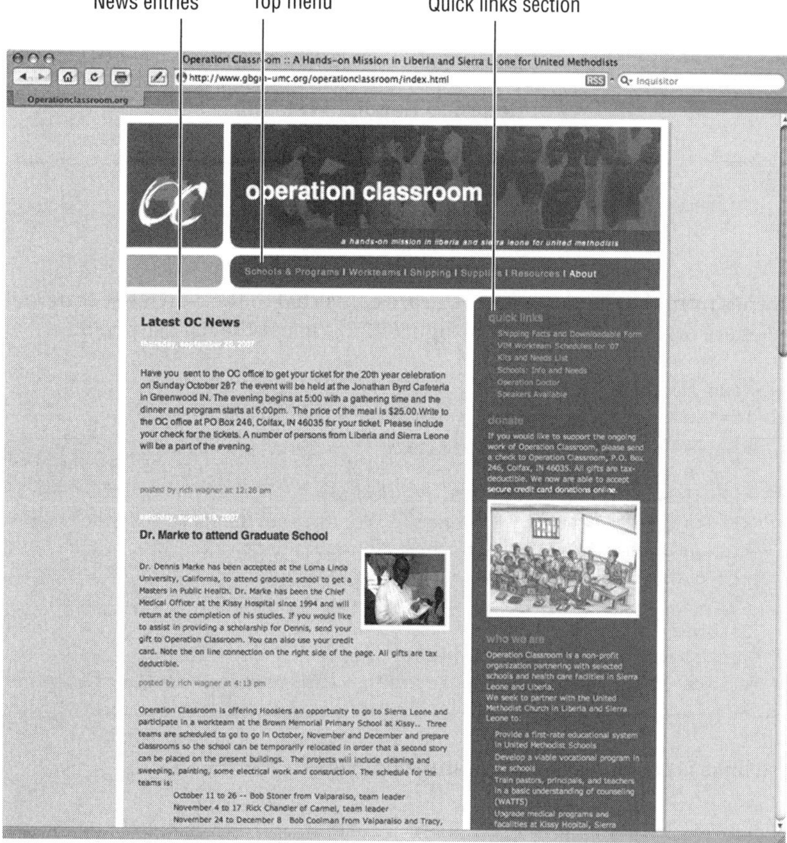

Figure 12-16: Operation Classroom home page

The top list items include both the top-level links from the regular site, along with news entries, quick links, and a link back to the regular Web site. Figure 12-17 shows the top-level menu when displayed on the iPhone.

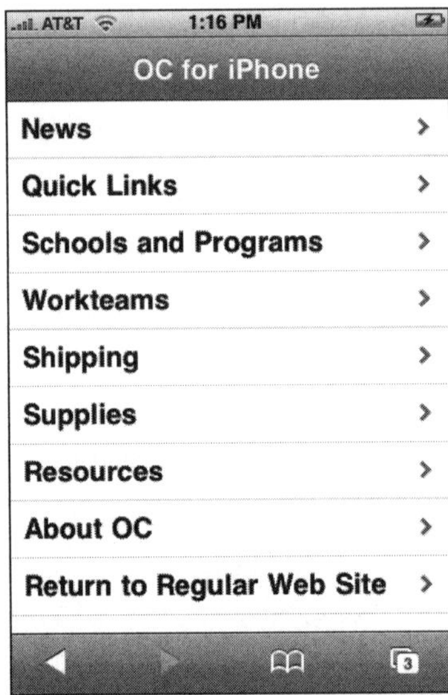

Figure 12-17: OC for iPhone

The news entries from the regular home page are converted to their own list of news articles. Notice that the entries are organized by date (see Figure 12-18) using the iUI class `group`:

```
<!--News menu-->
<ul id="news" title="Latest News">
    <li class="group">Sept. 20, 2009</li>
    <li><a href="#news1">20 Year Celebration Coming Soon</a></li>
    <li class="group">Aug. 18, 2009</li>
    <li><a href="#news2">Dr. Marke To Attend Graduate School</a></li>
    <li><a href="#news3">Workteam Scheduled for Kissy Clinic</a></li>
    <li class="group">June 23, 2009</li>
    <li><a href="#news4">Special Speakers Coming to Indiana in
    October</a></li>
    <li class="group">May 24, 2009</li>
    <li><a href="#Bill_Drake">Combat Malnutrition in Sierra Leone</a></li>
</ul>
```

Each of these links is connected with a destination page:

```
<div id="news1" class="panel" title="OC News">
<h2>20 Year Celebration Coming Soon</h2>
<p>Have you sent to the OC office to get your ticket for the 20 year
celebration on Sept. 20, 2009? The event will be held at JB's Cafeteria
in Greeley, IN. The evening begins at 5:00pm with a gathering time, and the
dinner and program start at 6:00pm. The price of the meal is $25.00. E-mail
the OC office for your ticket. Please include your check for the tickets. A
number of persons from Liberia and Sierra Leone will be a part of the
evening.</p>
</div>
```

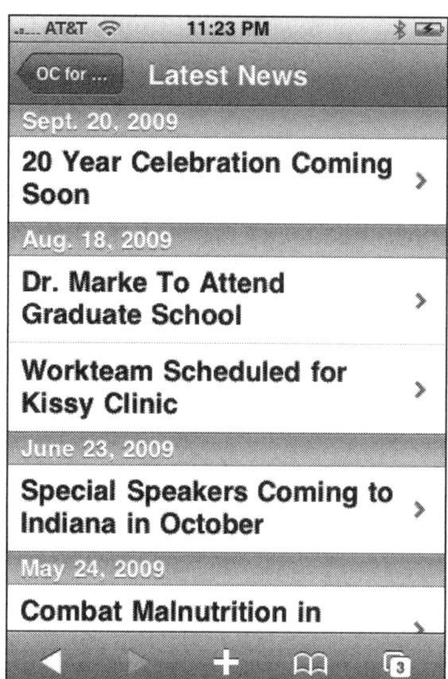

Figure 12-18: News entries by date

Figure 12-19 displays the results of this page.

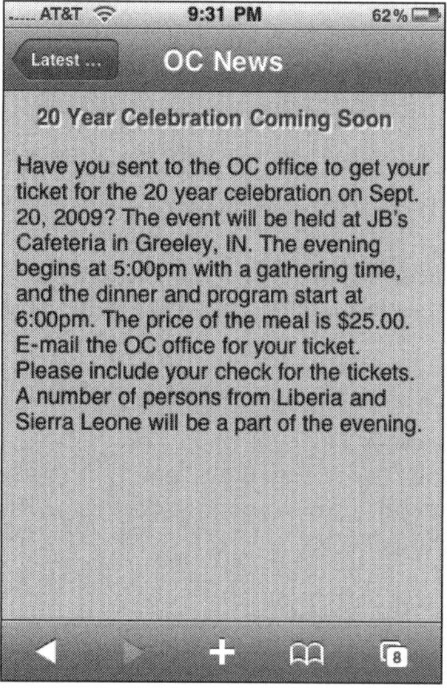

Figure 12-19: News article as a destination page

iPhone service integration offers you some special things with Contact Us pages. For example, when displaying contact information for the organization, you use a cUI destination page, which is discussed in Chapter 10:

```
<!--Contact panel-->
<div id="about" title="About Us" class="panel">
    <div class="cuiHeader">
        <img class="cui" src="images/iclass.png"/>
      <h1 class="cui">Operation Classroom</h1>
      <h2 class="cui">Partnering in Sierra Leone and Liberia</h2>
    </div>
    <fieldset>
        <div class="row">
            <label class="cui">office</label>
            <a class="cuiServiceLink" target="_self"
            href="tel:(765) 555-1212"
            >(765) 555-1212</a>
        </div>
        <div class="row">
            <label class="cui">mobile</label>
            <a class="cuiServiceLink" target="_self"
            href="tel:(765) 545-1211"
            >(765) 545-1211</a>
        </div>
        <div class="row">
            <label class="cui">email</label>
            <a class="cuiServiceLink" target="_self"
            href="mailto:info@operationclassroom.org"
            >info@oc.org</a>
        </div>
    </fieldset>
    <fieldset>
        <div class="rowCuiAddressBox">
            <label class="cui">office</label>
            <p class="cui">P.O. Box 120209.N</p>
            <p class="cui">Colfax, IN 46035</p>
        </div>
    </fieldset>
    <fieldset>
        <div class="row">
            <a  class="cuiServiceButton" target="_self"
            href="http://maps.google.com/maps?q=2012+Main,+Lapel,+IN">Map
            To Warehouse</a>
        </div>
    </fieldset>
</div>
```

The top `div` contains a thumbnail image and a place for company name and tagline. The next `fieldset` provides telephone and e-mail links (see Figure 12-20). iPod touch users will not be able to link to Phone or Mail applications, so an `click` handler is added to each link to enable the link only if it is running on an iPhone. (That's also why the text label for the e-mail link displays the actual address instead of "E-Mail Us.") The middle `fieldset` provides static address information, whereas the bottom `fieldset` contains a Google Maps link to the Operation Classroom warehouse (see Figure 12-21). If running on an iPhone, the Maps app opens. If running on an iPod touch, the Google Maps Web site is displayed.

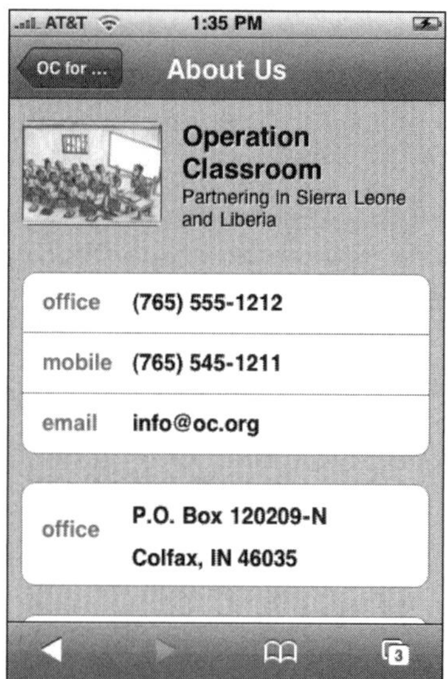

Figure 12-20: iPhone service integration

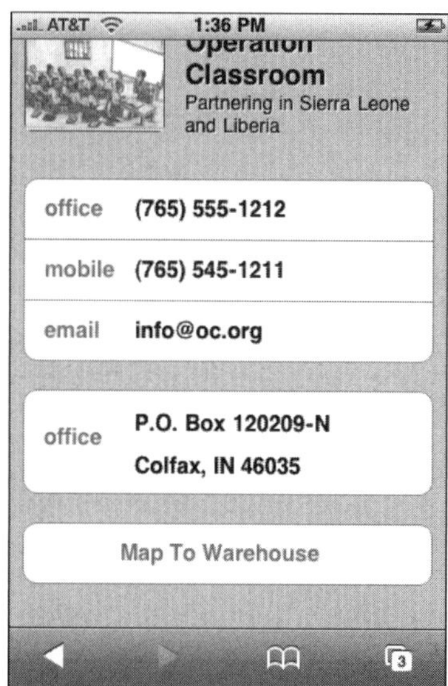

Figure 12-21: The link to Google Maps
works with both iPhone and iPod touch.

The following code shows the full source code for the sample OC-for-iPhone site. Note that many sections are not shown for this example that need to be implemented for an actual site.

```
<!DOCTYPE html PUBLIC "-//W3C//DTD XHTML 1.0 Strict//EN"
         "http://www.w3.org/TR/xhtml1/DTD/xhtml1-strict.dtd">
```

```html
<html xmlns="http://www.w3.org/1999/xhtml">
<head>
<title>Operation Classroom</title>
<meta name="viewport" content="width=320; initial-scale=1.0;
maximum-scale=1.0; user-scalable=0;"/>

<style type="text/css" media="screen">@import "./iui/iui.css";</style>
<style type="text/css" media="screen">@import "./iui/cui.css";</style>
<script type="application/x-javascript" src="./iui/iui.js"></script>
</head>
<body>
    <!--Top iUI toolbar-->
  <div class="toolbar">
      <h1 id="pageTitle"></h1>
      <a id="backButton" class="button" href="#"></a>
  </div>
  <!--Top-level menu-->
  <ul id="home" title="OC for iPhone" selected="true">
      <li><a href="#news">News</a></li>
      <li><a href="#quick-links">Quick Links</a></li>
      <li><a href="#schools-programs">Schools and Programs</a></li>
      <li><a href="#workteams">Workteams</a></li>
      <li><a href="#shipping">Shipping</a></li>
      <li><a href="#supplies">Supplies</a></li>
      <li><a href="#resources">Resources</a></li>
      <li><a href="#about">About OC</a></li>
      <li><a href="index.html" target="_self">Return to Regular web
      Site</a></li>
  </ul>
  <!--News menu-->
  <ul id="news" title="Latest News">
      <li class="group">Sept. 20, 2009</li>
      <li><a href="#news1">20 Year Celebration Coming Soon</a></li>
      <li class="group">Aug. 18, 2009</li>
      <li><a href="#news2">Dr. Marke To Attend Graduate School</a></li>
      <li><a href="#news3">Workteam Scheduled for Kissy Clinic</a></li>
      <li class="group">June 23, 2009</li>
      <li><a href="#news4">Special Speakers Coming to Indiana in
      October</a></li>
      <li class="group">May 24, 2009</li>
      <li><a href="#Bill_Drake">Combat Malnutrition in Sierra Leone</a></li>
  </ul>
  <div id="news1" class="panel" title="OC News">
    <h2>20 Year Celebration Coming Soon</h2>
    <p>
Have you sent to the OC office to get your ticket for the 20th year
celebration on Sept. 20, 2009Sunday October 28? The event will be held at
JB's Cafeteria in Greeley, IN. The evening begins at 5:00pm with a gathering
time, and the dinner and program starts at 6:00pm. The price of the meal is
$25.00. E-mail the OC office for your ticket. Please include your check for
the tickets. A number of persons from Liberia and Sierra Leone will be a part
of the evening.</p>
  </div>
  <!--More content would appear here-->
```

```html
<!--About Us panel-->
<div id="about" title="About Us" class="panel">
    <div class="cuiHeader">
        <img class="cui" src="images/iclass.png"/>
    <h1 class="cui">Operation Classroom</h1>
    <h2 class="cui">Partnering in Sierra Leone and Liberia</h2>
  </div>
  <fieldset>
      <div class="row">
          <label class="cui">office</label>
          <a class="cuiServiceLink" target="_self"
          href="tel:(765) 555-1212"
          >(765) 555-1212</a>
      </div>
      <div class="row">
          <label class="cui">mobile</label>
          <a class="cuiServiceLink" target="_self"
          href="tel:(765) 545-1211"
          >(765) 545-1211</a>
      </div>
      <div class="row">
          <label class="cui">email</label>
          <a class="cuiServiceLink" target="_self"
          href="mailto:info@operationclassroom.org"
          >info@oc.org</a>
      </div>
  </fieldset>
  <fieldset>
      <div class="rowCuiAddressBox">
          <label class="cui">office</label>
          <p class="cui">P.O. Box 120209.N</p>
          <p class="cui">Colfax, IN 46035</p>
      </div>
  </fieldset>
  <fieldset>
    <div class="row">
        <a  class="cuiServiceButton" target="_self"
        href="http://maps.google.com/maps?q=2012+Main,+Lapel,+IN">Map
        To Warehouse</a>
    </div>
  </fieldset>
</div>
</body>
</html>
```

Summary

This chapter concentrates on the issues that you'll want to consider when you provide support for iPhone and iPod touch on your regular Web site. You need to determine the level of support you wish to offer. I focus on four tiers: simple compatibility, device friendliness, iPhone-specific styling, and dedicated companion Web site. In doing so, I focus on taking a single Web site and walking you through how that site might look with each of these levels of support.

13

Bandwidth and Performance Optimizations

In a decade where broadband is now the norm, many Web developers have fallen into those same tendencies and allow their sites and applications to be composed of ill-formed HTML, massive JavaScript libraries, and multiple CSS style sheets.

However, when you are developing applications for iPhone and iPod touch, you need to refocus your programming and development efforts toward optimization and efficiency. What makes them different from normal Web apps is that the developer can no longer rely on the fact that users are accessing the application from a broadband connection. iPhone users may be coming to your application using Wi-Fi or a slower 3G connection.

Therefore, as you develop your applications, you will want to formulate an optimization strategy that makes the most sense for your context. You'll want to think about both bandwidth and code performance optimizations.

Optimization Strategies

If you spend much time at all researching optimization strategies and techniques, you quickly discover two main schools of thought. The first camp is referred to as *hyper-optimizers* in this book. They will do almost anything to save a byte or an unneeded call to the Web server. They are far more concerned with saving milliseconds than they are about the readability of the code they are optimizing. The second camp, perhaps best described as *relaxed optimizers*, are interested in optimizing their applications. But they are not interested in sacrificing code readability and manageability in an effort to save a nanosecond here or there.

Decide which camp you fall into. At the same time, though, don't go through complex optimization hoops unless you prove that your steps are going to make a substantive difference in the usability of your application. Many optimization techniques you'll find people advocating may merely make your code harder to work with and not offer notable performance boost.

Best Practices to Minimize Bandwidth

Arguably the greatest bottleneck of any iPhone Web app is the time it takes to transport data from the Web server to Safari, especially if your application is running over 3G. You can certainly opt for local offline caching (see Chapter 11, "Offline Applications"). However, you should also consider the following techniques as you assemble your Web application.

General

- ❑ Separate your page content into separate CSS, JS, and HTML files so that each file can be cached by Safari on iPhone.

- ❑ Meanwhile, studies show that loading additional files takes up 60–90 percent of the total time of a Web page. Therefore, be sure to minimize the total number of external style sheets, JavaScript library files, and images you include with your page. Because browsers typically make two to four requests at a given time, every additional file that a browser has to wait on for the request to complete will create latency.

- ❑ Reduce whitespace (tabs and spaces) wherever possible. Although this might seem like a nominal issue, the amount of excess whitespace can add up, particularly on a larger-scale Web application with dozens of files.

- ❑ Remove useless tags, and unused styles and JavaScript functions in your HTML, CSS style sheets, and JavaScript library files.

- ❑ Remove unnecessary comments. However, keep in mind the following caveat: removing comments can reduce file size, but it can also make it harder to manage your code in the future.

- ❑ Use shorter filenames. For example, it is much more efficient to reference `tb2.png` than `TopBannerAlternate2_980.png`.

- ❑ Write well-formed and standard XHTML code. Although not a bandwidth issue, well-formed XHTML requires fewer passes and parsing by Safari before it renders the page. As a result, the time from initial request to final display can be improved through this coding practice.

- ❑ Consider using gzip compression when you serve your application. (See the following section for more on compression options.)

- ❑ Consider using a JavaScript compressor on your JavaScript libraries. You could then work with a normal, unoptimized JavaScript library for development (`mylibrary.js`) and output a compressed version for runtime purposes (`mylibrary-c.js`). (See the following section for more on compression options.)

Images

- ❑ Large image sizes are a traditional bottleneck to target for your applications. Be meticulous in optimizing the file size of your images. Shaving off 5Kb or so from several images in your application can make a notable performance increase.

- ❑ Make sure your images are sized appropriately for display on the iPhone viewport. Never rely on browser scaling. Instead, match image size to image presentation.

❑ Image data is more expensive than text data. Therefore, consider using a canvas drawing in certain cases.

❑ Instead of using image borders, consider using CSS borders, particularly with the enhanced `-webkit-border-radius` property.

❑ Instead of using one large background image, consider using a small image and tiling it.

❑ Use CSS sprites to combine multiple image files into a larger one. The overall latency impact is reduced because fewer requests are made to the server.

CSS and JavaScript

❑ Combine rules to create more efficient style declarations. For example, the second declaration is much more space efficient than the first one is:

```
// Less efficient
div #content {
                font-family: Helvetica, Arial, sans-serif;
                font-size: 12px; /* Randy: do we want this as px or pt? */
                line-height: 1.2em; /* Let's try this for now.*/
                font-weight: bold;
}
// More efficient
div #content {font: bold 12px/1.2em Helvetica, Arial, sans-serif};
```

❑ Consider using shorter CSS style names and JavaScript variable and function names. After all, the longer your identifiers are, the more space your files will take. But, at the same time, do not make your identifiers so short that they become hard to work with. For example, consider the trade-offs with the following three declarations:

```
/* Inefficient */
#homepage-blog-subtitle-alternate-version{letter-spacing:.1em;}
/* Efficient, but cryptic */
#hbsa{letter-spacing:.1em;}
/* Happy medium */
#blog-subtitle-alt{letter-spacing:.1em;}
```

As you work through these various strategies and test results, a good way to check the total page size is to save the page as a Web archive in a desktop version of Safari. The size of the archive file indicates the HTML page size with all the external resources (images, style sheets, and script libraries) associated with it.

Compressing Your Application

Normally, an iPhone Web application is launched when users type the URL in their Safari browser (or click its shortcut icon on the iPhone home screen). The Web server responds to the HTTP request and serves the HTML file and each of the many supporting files that are used in the display and execution of the Web app. Although image files may have been optimized as much as possible to minimize bandwidth, each uncompressed HTML file, CSS style sheet, and JavaScript library file requested takes up much more space than if it were compressed. With that idea in mind, several options are available to compress files or JavaScript code on the fly on the server.

Gzip File Compression

Safari supports gzip compression, which many Web servers offer. Using gzip compression, you can reduce the size of HTML, CSS, and JavaScript files and reduce the total download size by up to 4 to 5 times. However, because Safari must uncompress the resources when it receives them, be sure to test that this overhead does not eliminate the benefits gained.

For example, if you wanted to turn on gzip compression in PHP, you could use the following code:

```
<?php
ob_start("ob_gzhandler");
?>
<html>
<body>
<p>This page has been compressed.</p>
</body>
</html>
```

JavaScript Code Compression

In addition to reducing the total file size of your Web site, another technique is to focus on JavaScript code. These compression strategies go far beyond the manual coding techniques described in this chapter and seek to compress and remove all unnecessary characters in your JavaScript code. In fact, using these automated solutions, you can potentially reduce the size of your scripts by 30–40 percent.

You can turn to a variety of open source solutions, which tend to take two different approaches. The safe optimizers remove whitespace and comments from code but do not seek to change naming inside of your source code. The hyper-optimizers go a step further and seek to crunch variable and function names. While the hyper-optimizers achieve greater compression ratios, these ratios are not as safe to use in certain situations. For example, if you have `eval()` or `with` in your code (not recommended anyway), they are broken during the compression process. What's more, some of the optimizers, such as Packer, use an `eval`-based approach to compress and uncompress. However, there is a performance hit in the decompression process, and it could actually slow down your script under certain conditions.

Here are some of the options available (ranked in order of conservatism employed in their algorithms):

- ❑ JSMin (JavaScript Minifier; `www.crockford.com/javascript/jsmin.html`) is perhaps the best-known JavaScript optimizer. It is the most conservative of the optimizers, focusing on simply removing whitespace and comments from JavaScript code.

- ❑ YUI Compressor (`http://developer.yahoo.com/yui/compressor/`) is an optimizer that claims to offer a happy medium between the conservative JSMin and the more aggressive ShrinkSafe and Packer listed next.

- ❑ Dojo ShrinkSafe (`www.dojotoolkit.org/docs/shrinksafe`) optimizes and crunches local variable names to achieve greater compression ratios.

- ❑ Dean Edwards's Packer (`dean.edwards.name/packer/`) is a hyper-optimizer that achieves high compression ratios.

Deciding which of these options to use should depend on your specific needs and the nature of your source code. I recommend starting on the safe side and moving up as needed.

If you decide to use one of these optimizers, make sure you use semicolons to end your lines in your source code. Besides being good programming practice, most optimizers need these punctuation marks to accurately remove excess whitespace.

Additionally, whereas Packer requires semicolons, Dojo ShrinkSafe does not require them and actually inserts missing semicolons for you. So you can preprocess a JavaScript file through ShrinkSafe before using it in a semicolon-requiring compressor like Packer.

To demonstrate the compression ratios that you can achieve, I ran the `iUI.js` JavaScript library file through several of these optimizing tools. Table 13-1 displays the results.

Table 13-1: Benchmark of Compression of iUI.js File

Compressor	JavaScript Compression (Bytes)	With gzip Compression (Bytes)
No compression	100% (11284)	26% (2879)
JSMin	65% (7326)	21% (2403)
Dojo ShrinkSafe	58% (6594)	21% (2349)
YUI Compressor	64% (7211)	21% (2377)
YUI Compressor (w/Munged)	46% (5199)	18% (2012)
YUI Compressor (w/Preserve All Semicolons)	64% (7277)	21% (2389)
YUI Compressor (w/Munged and Preserve All Semicolons)	47% (5265)	18% (2020)

One final option worth considering is a PHP-based open source project called Minify. Minify combines, minifies, and caches JavaScript and CSS files to decrease the number of page requests that a page has to make. To do so, it combines multiple style sheets and script libraries into a single download (`code .google.com/p/minify/`).

JavaScript Performance Optimizations

The performance of JavaScript on iPhone is slower than on the Safari desktop counterparts, although recent improvements in the iPhone 3.0 OS shrink this gap. For example, consider the following simple DOM-access performance test:

```
<!DOCTYPE html PUBLIC "-//W3C//DTD XHTML 1.0 Strict//EN"
        "http://www.w3.org/TR/xhtml1/DTD/xhtml1-strict.dtd">
<html xmlns="http://www.w3.org/1999/xhtml">
```

```html
<head>
<title>Performance Test</title>
</head>
<body>
<form id="form1">
<input id="i1" value="zero" type="text">
</form>
<div id="output"></div>
</body>
<script type="text/javascript" language="javascript">
var i = 0;
var start1 = new Date().getTime();
divs = document.getElementsByTagName('div');
for(i = 0; i < 80000; i++) {
  var d = divs[0];
}
var start2 = new Date().getTime();
var delta1 = start2--start1;
document.getElementById("output").innerHTML = "Time: " + delta1;
</script>
</html>
```

Safari for Mac OS X executes this script in 529 milliseconds, whereas Safari for iPhone takes 13,922 milliseconds. That's more than 26 times longer! Therefore, in addition to the optimizations you can make to shrink the overall file size of your application, you should give priority to making performance gains in execution based on your coding techniques. Here are several best practices to consider.

Smart DOM Access

When working with client-side JavaScript, accessing the DOM can be at the heart of almost anything you do. However, as essential as these DOM calls may be, it is important to remember that DOM access is expensive from a performance standpoint and should be done with forethought.

Cache DOM References

Cache references that you make to avoid multiple lookups on the same object or property. For example, compare the following inefficient and efficient routines:

```javascript
// Ineffecient
var str = document.createTextNode("Farther up, further in");
document.getElementById("para1").appendChild(str);
document.getElementById("para1").className="special";
// More efficient
var str = document.createTextNode("Farther up, further in");
var p = document.getElementById("para1");
p.appendChild(str);
p.className="special";
```

What's more, if you make a sizeable number of references to a document or another common DOM object, cache them, too. For example, compare the following:

```javascript
// Less efficient
var l1=document.createTextNode('Line 1');
```

```
var l2=document.createTextNode('Line 2');
// More efficient
var d=document;
var l1=d.createTextNode('Line 1');
var l2=d.createTextNode('Line 2');
```

If you reference `document` a handful of times, it is probably not practical to go through this trouble. But if you find yourself writing `document` a thousand times in your code, the efficiency gains make this practice a definite consideration.

Offline DOM Manipulation

When you are writing to the DOM, assemble your subtree of nodes outside of the actual DOM, and then insert the subtree once at the end of the process. For example, consider the following:

```
var comments=customBlog.getComments('index');
var c=comments.count;
var entry;
var commentDiv = document.createElement('div');
document.body.appendChild(commentDiv);
for (var i = 0; i < c; i++) {
  entry=document.createElement('p');
  entry.appendChild( document.createTextNode(comments[i]);
  commentDiv.appendChild(entry);
}
```

Consider the placement of the grayed, highlighted line. Because you add the new `div` element to the DOM before you add children to it, the document must be updated for each new paragraph added. However, you can speed up the routine considerably by moving the offending line to the end:

```
var comments=customBlog.getComments('index');
var c=comments.count;
var entry;
var commentDiv = document.createElement('div');
for (var i = 0; i < c; i++) {
  entry=document.createElement('p');
  entry.appendChild( document.createTextNode(comments[i]);
  commentDiv.appendChild(entry);
}
document.body.appendChild(commentDiv);
```

With the restructured code, you need to update the document display only once instead of multiple times.

Combining document.write() Calls

Along the same line, you should avoid excessive `document.write()` calls because each call is a performance hit. A much better practice is to assemble a concatenated string variable first. For example, compare the following:

```
// Inefficient
document.write('<div class="row">');
document.write('<label class="cui">office</label>');
```

```
document.write('<a class="cuiServiceLink" target="_self" href="tel:(765) 555-1212">
(765) 555-1212</a>');
document.write('</div>');
// More efficient
var s = '<div class="row">' + '<label class="cui">office</label>' +
'<a class="cuiServiceLink" target="_self" href="tel:(765) 555-1212">(765) 555-1212
</a>' + '</div>';
document.write(s);
```

Using the Window Object

The window object is faster to use because Safari does not have to navigate the DOM to respond to your call. The following window reference is more efficient than the previous three approaches:

```
// Inefficient
var h=document.location.href;
var h=document.URL;
var h=location.href;
// More efficient
var h=window.location.href
```

Local and Global Variables

One of the most important practices JavaScript coders should implement in their code is use of local variables and avoidance of global variables. When Safari processes a script, local variables are looked for first in the local scope. If Safari can't find a match, it moves up the next level, and then the next, until it hits the global scope. So global variables are the slowest in a lookup. For example, defining variable a at the global level in the following code is much more expensive than defining it as a local variable inside of the for routine:

```
// Inefficient
var a=1;
function myFunction(){
  for(var i = 0; i < 10; i++) {
    var t = a+i;
    // do something with t
  }
}
//More efficient
function myFunction(){
  for(var i = 0, a = 1; i < 10; i++) {
    var t = a+i;
    // do something with t
  }
}
```

Dot Notation and Property Lookups

Accessing objects and properties by dot notation is never efficient. Therefore, consider some alternatives.

Avoiding Nested Properties

Aim to keep the levels of dot hierarchy small. Nested properties, such as `document.property.property.property`, cause the biggest performance problems and should be avoided or accessed as few times as possible.

```
// Inefficient
m.n.o.p.doThis();
m.n.o.p.doThat();
// More efficient
var d = m.n.o.p;
d.doThis();
d.doThat();
```

Accessing a Named Object

If you access a named object, it is more efficient to use `getElementById()` than to access it via dot notation. For example, compare the following:

```
// Inefficient
document.form1.addressLine1.value
// More efficient
document.getElementById('addressLine1').value;
```

Property Lookups Inside Loops

When accessing a property inside a loop, it is much better practice to cache the property reference first and then access the variable inside the loop. For example, compare the following:

```
// Inefficient
for(i = 0; i < 10; i++) {
var v = document.object.property(i);
var y = myCustomObject.property(i);
// do something
}
// More efficient
var p = document.object.property;
var cp = myCustomObject.property(i);
for(i = 0; i < 10; i++) {
var v= p(i);
var y=cp(i);
// do something
}
```

Here's another example of using the `length` property of an object in the condition of a `for` loop:

```javascript
// Inefficient
for (i = 0; i < myObject.length; i++) {
  // Do something
}
// More efficient
var j = myObject.length;
for (i = 0,; i < j; i++) {
  // Do something
}
```

Similarly, if you are using arrays inside of loops and using its `length` as a conditional, you want to assign its length to a variable rather than evaluating at each pass. Check this out:

```javascript
// Inefficient
myArray = new Array();
for (i = 0; i < myArray.length; i++) {
  // Do something
}
// More efficient
myArray = new Array();
len = myArray.length;
for (i = 0; i < len; i++) {
  // Do something
}
```

String Concatenation

Another traditional problem area in JavaScript is string concatenation. In general, you should try to avoid an excessive number of concatenations and an excessively large string that you are appending to. For example, suppose you are trying to construct a table in code and then write out the code to the document when you are finished. The `stringTable()` function in the following code is less efficient than the second function `intermStringTable()`, because the latter uses an intermediate string variable `row` as a buffer in the `for` loop.

```html
<html>
<script type="text/javascript" language="javascript">
function stringTable() {
  var start = new Date().getTime();
  var buf = "<table>";
  for (var i = 0; i < 10000; i++){
    buf += "<tr>";
    for (var j = 0; j < 40; j++){
      buf += "<td><i>" + "content" + "</i></td>";
    }
    buf += "</tr>";
  }
  buf += "</table>";
  var duration = new Date().getTime() -- start;
  document.write('String concat method: ' + duration + '</br>');
}
function intermStringTable(){
  var start = new Date().getTime();
```

```
        var buf = "<table>";
      for (var i = 0; i < 10000; i++){
        var row = "<tr>";
        for (var j = 0; j < 40; j++){
          row += "<td><i>" + "content" + "</i></td>";
        }
        row += "</tr>";
        buf += row
      }
      buf += "</table>";
      var duration = new Date().getTime() -- start;
      document.write('Intermediate concat method: ' + duration + '</br>');
    }
    </script>
    <body>
    </body>
    <script type="text/javascript" language="javascript">
    stringTable();
    intermStringTable();
    </script>
    </html>
```

What to Do and Not to Do

Be sure to avoid `with` statements, which slow the processing of the related code block. Besides the fact that `with` is inefficient, it has been deprecated in the JavaScript standard. Also, avoid using `eval()` in your scripts because it is expensive from a performance standpoint. Besides, you should be able to develop a more efficient solution than resorting to `eval()`.

Comments add to readability and manageability, but be wise in their usage. For example, minimize their use inside of loop routines, functions, and arrays. If possible, place comments before or after a programming construct to ensure greater efficiency.

```
    // Inefficient
    var a = 0, c = 100;
    for (var i = 0; i < c; i++) {
      // Assign d the value of the next div in the current document
      var d = document.getElementByTagName('div')[i];
      // Perform some math for a
      a=i * 1.2;
      // Perform some math for b
      b=(a + i) / 3;
    }
    // More efficient
    // Assign val of d to 100 divs and perform y on them
    // based on val of a and b.
    var a = 0, c = 100;
    for (var i = 0; i < c; i++) {
      var d = document.getElementByTagName('div')[i];
      a-i * 1.2;
      b=(a + i) / 3;
    }
```

Summary

In this chapter, you focused on how to optimize your Web sites and apps. The great bottleneck of any iPhone Web app is the time it takes to move data from the Web server to the device. You will therefore want to separate your page content into separate files so that each file can be cached. At the same time, you will want to minimize the total number of these files. Also, focus considerable attention on image optimization. Large image sizes are a one of the greatest problems to deal with and so you will want to work hard to optimize the file size of your images.

A second area to focus on when optimizing your Web app is compression. Beyond image compression, consider compressing CSS style sheet and JavaScript library files. Using gzip compression, you can significantly reduce the download size of your Web app, as much as five times. What's more, consider using a JavaScript code compressor to shrink the size of your scripts by 40 percent.

Finally, in your JavaScript code, be sure to consider how to interact with the DOM of the page you are working with. You'll want to make use of variables and cache references that you make to avoid redundant lookups on the same DOM object. What's more, when you write to the DOM, be careful to minimize the number of successive `document.write()` calls. Combine when possible. Also, you'll want to assemble your subtree of nodes outside of the actual DOM, and then insert the subtree once at the end of the process. Finally, when possible, work with the window object instead of the document object or its children. The window object is faster and more efficient to call.

14

Packaging Apps as Bookmarks: Bookmarklets and Data URLs

Because iPhone Web applications function inside the Safari environment, as a Web developer, you have two seemingly obvious restrictions: you must live with the built-in capabilities of the Safari/WebKit browser; and you need a constant Wi-Fi (or, for iPhone, 3G) connection to run any application.

As you already found out in Chapter 13, "Bandwidth and Performance Optimizations," you can use HTML5 offline storage capabilities to get around the dependency on a live connection. However, two lesser-known technologies — bookmarklets and data URLs — can provide similar results. These technologies have actually been around for years, but they have tended to exist on the periphery of mainstream Web development. However, developers are now reexamining these two developer tools to maximize the potential of the iPhone Web application platform.

Bookmarklets (short for *bookmark applets*) are mini JavaScript "applets" that can be stored as bookmarks inside Safari. A data URL is a technique for storing an entire Web page or application (pages, styles, images, data, and scripts) inside a single URL, which can then be saved as an iPhone bookmark inside Safari. This application-in-a-bookmark can then be accessed in offline mode.

Bookmarklets

A *bookmarklet* is JavaScript stored as a URL and saved as a bookmark in the browser. It is typically used as a one-click applet that performs a specific task or performs an action on the current Web page. A bookmarklet uses the `javascript:` protocol followed by script code. For instance, here's the simplest of examples:

```
javascript:alert('iPhone')
```

Because the scripting code for a bookmarklet is housed inside a URL, the script must be condensed into one long string of code. Therefore, to enter multiple statements, separate each line with a semicolon:

```
javascript:alert('Bookmarklet 1');alert('Bookmarklet 2')
```

In this case, I added spaces inside each of the strings. You can either substitute %20 for a blank space or let Safari do the conversion for you.

If the script returns a value, be sure to enclose it inside of `void()` to ensure that the JavaScript code runs as expected. For example, the following WikiSearch bookmarklet displays a JavaScript prompt dialog box (see Figure 14-1) and then calls a Wikipedia search URL using the user's value as the search term:

```
javascript:t=prompt('Search Wikipedia:',getSelection());if(t)void(
location.href='http://en.wikipedia.org/w/wiki.phtml?search='+escape(t))
```

Figure 14-1: WikiSearch bookmarklet

Here's a second example that provides a front end onto Google's `define` service:

```
javascript:d=prompt('Define:',getSelection());if(d)void(
location.href='http://www.google.com/search?q=define:'+escape(d))
```

Adding a Bookmarklet to Safari for iPhone

Bookmarklets are normally added in a standard browser through a drag-and-drop action. However, because that user input is not available in Safari on iPhone, you need to add the bookmarklet through the following process:

1. On your main computer, create your bookmarklet script and test it by pasting it into the Address box of Safari.

2. Once the functionality works as expected, drag the `javascript:` URL from your Address box onto your Bookmarks bar in Safari. If you are going to have a set of bookmarklets, you may want to create a special Bookmarklets folder to store these scripts.

   Or, if your bookmarklet is contained within the `href` of an `a` link, drag the link onto the Bookmarks bar instead.

3. Synch the bookmarks of your iPhone and main computer through iTunes.

4. Access the bookmarklet in the Bookmarks inside Safari (see Figure 14-2).

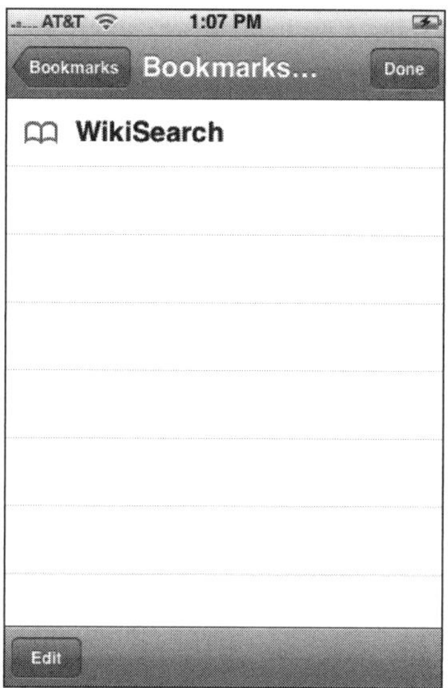

Figure 14-2: Accessing a bookmarklet from iPhone

Alternatively, you can add a bookmarklet directly into Safari's Bookmarks by creating a link to any normal Web page and then editing the URL of the bookmark.

Exploring How Bookmarklets Can Be Used

Although bookmarklets can be used for these sorts of general purposes, their real usefulness to the iPhone Web app developer is turning JavaScript into a macro language for Safari to extend the functionality of the browser. For example, Safari always opens normal links in the existing window, replacing the existing page. Richard Herrera from `www.doctyper.com` wrote a bookmarklet that transforms the links of a page and forces them to open in a new tab. Here is the script, which is tricky to read because it is contained within a one-line, encoded URL:

```
javascript:(function(){var%20a=document.getElementsByTagName('a');for(var%20i=0,
j=a.length;i%3Cj;i++){a[i].setAttribute('target','_blank');var%20img=document.
createElement('img');img.setAttribute('class',%20'new-window');img.setAttribute
('src','data:image/gif;base64,'+'R0lGODlhEAAMALMLAL66tBISEjFxMdTQyBoaGjs7OyUlJ
WZmZgAAAMzMzP///////wAAAAAAAAAAAAA'+'ACH5BAEAAAsALAAAAAAQAAwAAAQ/cMlZqr2Tps13
yVJBjOT4gYairqohCTDMsu4iHHgwr7UA/LqdopZS'+'DBBIpGG5lBQH0GgtU9xNJ9XZ1cnsNicRADs
=');img.setAttribute('style','width:16px!important;height:12px!important;
border:none!important;');a[i].appendChild(img);}})();
```

When executed, the script adds a `target="_blank"` attribute to all links on the page and adds a small "new window" icon after the link (see Figure 14-3).

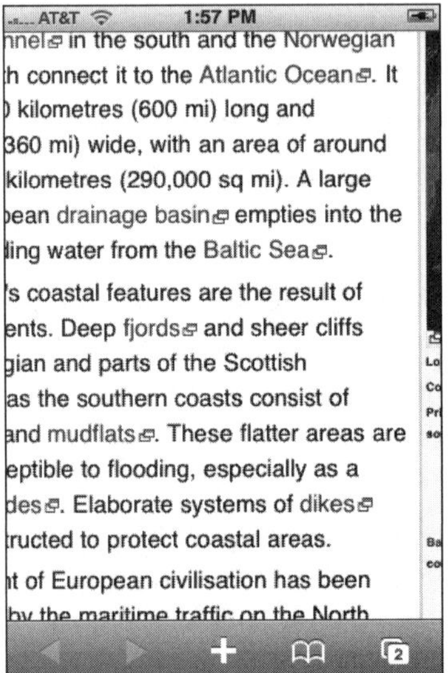

Figure 14-3: New window icons added after links

iPhone users can then use this self-contained "applet" on any page in which they want to transform the links. Notice that the icon image used in the script is encoded inside a data URL, so the script does not depend on external files.

Although the entire script needs to be condensed into a single string of commands, Safari is actually smart enough to convert the hard breaks for you when a multilined script is pasted into the URL box. Just make sure each statement is separated by a semicolon. Therefore, the following code, which is much easier to work with and debug, would still execute properly when pasted directly into the URL box:

```
javascript:(
 function(){
        var a=document.getElementsByTagName('a');
        for(var i=0,j=a.length;i%3Cj;i++) {
                a[i].setAttribute('target','_blank');
                var img=document.createElement('img');
                img.setAttribute('class','new-window');
  img.setAttribute('src','data:image/gif;base64,'+'R0lGODlhEAAMALMLAL66tBISEjExMdTQ
yBoaGjs7OyUlJWZmZgAAAMzMzP///////wAAAAAAAAAAAA'+'ACH5BAEAAAsALAAAAAAQAAwAAAQ/cM1
Zqr2Tps13yVJBjOT4gYairqohCTDMsu4iHHgwr7UA/LqdopZS'+'DBBIpGG5lBQH0GgtU9xNJ9XZ1cnsNi
cRADs=');
                img.setAttribute('style','width:16px!important;
                height:12px!important;
                border:none!important;');
                a[i].appendChild(img);
        }
 })();
```

Bookmarklets can be handy developer tools to assist in testing and debugging on iPhone. For example, the following bookmarklet, based on a script created at www.iPhoneWebDev.com, gives you View Source functionality (see Figure 14-4) on iPhone:

```
javascript:
var sourceWindow = window.open("about:blank");
var newDoc = sourceWindow.document;
newDoc.open();
newDoc.write(
"<html><head><title>Source of " + document.location.href +
"</title><meta name=\"viewport\" id=\"viewport\" content=\"initial-scale=1.0;" +
"user-scalable=0;maximum-scale=0.6667;width=480\"/><script>function do_onload()"
+ "{setTimeout(function(){window.scrollTo(0,1);},100);}if(navigator.userAgent.
indexOf" + "(\"iPhone\")!=-1)window.onload=do_onload;</script></head><body>
</body></html>");
newDoc.close();
var pre = newDoc.body.appendChild(newDoc.createElement("pre")); pre.appendChild(
newDoc.createTextNode(document.documentElement.innerHTML));
```

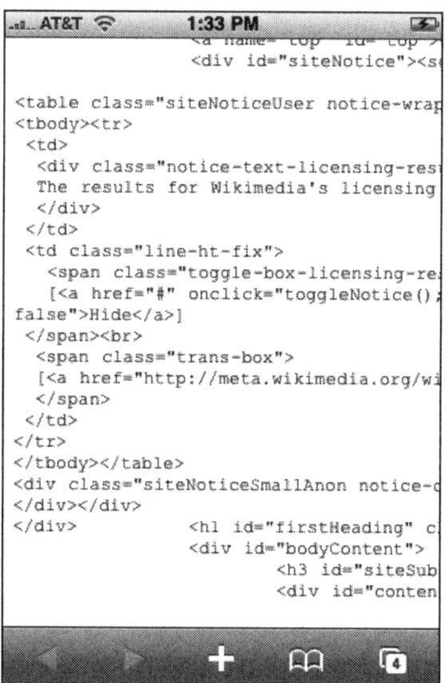

Figure 14-4: Viewing a page's source on iPhone

If you'd like to work with these sample bookmarklets, save yourself some typing time by going to www.wrox.com and downloading these instead.

Storing an Application in a Data URL

In addition to JavaScript functionality, you can store a Web page or even a complete application inside of a bookmark. The data: protocol allows you to encode an entire page's content — HTML, CSS, JavaScript, and images — inside a single URL. To be clear, data URLs store not a simple link to a remote page, but the actual contents of the page. This data URL can then be saved as a bookmark. When users access this bookmark in Safari, they can interact with the page whether or not they have Internet access. The implications are significant — you can use data URLs to package certain types of Web applications and get around the live Internet connection requirement.

Constraints and Issues with Using Data URLs

Although the potential of data URLs is exciting for the developer, make sure you keep the following constraints and issues in mind before working with them:

❑ You can store client-side technologies — such as HTML, CSS, JavaScript, and XML — inside a data URL. However, you *cannot* package PHP, MySQL, or any server-side applications in a bookmark.

❑ Any Web application that requires server access for data or application functionality needs to have a way to pack and go: (1) use client-side JavaScript for application functionality, and (2) package a snapshot of the data and put it in a form accessible from a client script. However, in most cases in which you need to use cached data, you'll want to use HTML5 offline storage instead (see Chapter 11).

❑ The Web application must be *entirely* self-contained. Therefore, every external resource the application needs, such as images, style sheets, and .js libraries, must be encoded inside the main HTML file.

❑ External resources that are referenced multiple times cannot be cached. Therefore, each separate reference must be encoded and embedded in the file.

❑ Images must be encoded as Base64, although the conversion increases their size by approximately 33 percent. (*Base64* is the process of encoding binary data so it can be represented with normal character set characters. Encoded Base64 data must then be decoded before it can be used.)

❑ The maximum size of a data URL in Safari for iPhone is technically 128KB, although in actual practice, you can work with URLs much larger, at least up to several megabytes. However, performance of the Safari Bookmark manager suffers significantly when large amounts of data are stored inside a bookmark. Therefore, think thin for data URL–based applications.

❑ Safari has issues working with complex JavaScript routines embedded in a data URL application. For example, the UI frameworks discussed in Chapter 3, "Building with Web App Frameworks," may not functional inside a data URL, thus greatly limiting the potential for Web developers to take advantage of this offline storage option.

❑ If your development computer is a Mac, you should be okay working with data URLs. However, if you are working with the Windows version of Safari and trying to synch the bookmark with Safari for iPhone, be careful: Safari for Windows has major limitations in the size of data it can store inside a bookmark. Consider using HTML5 offline cache (see Chapter 11) instead.

Creating a Data URL App

After examining these constraints, it is clear that the best candidates for data URL apps are those that are relatively small in both scope and overall code base. A tip calculator, for example, is a good sample applet because its UI would be simple and its programming logic would be straightforward and not require accessing complex JavaScript libraries. I'll walk you through the steps needed to create a data URL application.

After reviewing the constraints and making sure that your application will likely work in an offline mode, you will want to begin by designing and programming as if it were a normal iPhone Web app application. For this sample applet, the interface of the tip calculator is based on a subset of a legacy version of the iUI framework. (To limit the size of the app, I am not including references to the framework.)

Figure 14-5 shows the Tipster application interface that you will be constructing.

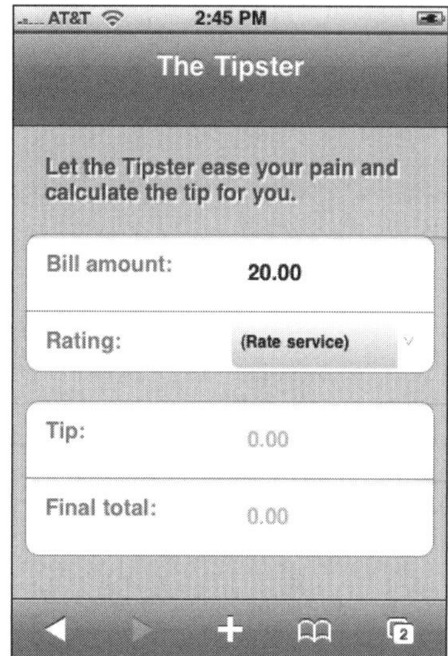

Figure 14-5: Tipster application design

The following source file shows the core HTML and JavaScript code:

```
<!DOCTYPE html PUBLIC "-//W3C//DTD XHTML 1.0 Strict//EN"
        "http://www.w3.org/TR/xhtml1/DTD/xhtml1-strict.dtd">
<html xmlns="http://www.w3.org/1999/xhtml">
<head>
<title>Tipster</title>
<meta name="viewport" content="width=320; initial-scale=1.0; maximum-scale=1.0;
user-scalable=0;"/>
<script type="text/javascript" language="javascript">

addEventListener('load', function()
{
  setTimeout(function() {
      window.scrollTo(0, 1);
  }, 100);
 }, false);

function checkTotal(fld)
{
```

```
        var x=fld.value;
        var n=/(^\d+$)|(^\d+\.\d+$)/;
        if (n.test(x))
        {
            if (fldTipPercent.selectedIndex != 0) getRec();
        }
        else
        {
            alert('Please enter a valid total')
            clearTotal(fld);
        }
}

function clearTotal(fld)
{
  fld.value = '';
}

function getRec()
{
    if (fldTipPercent.selectedIndex == 0)
    {
        alert('Please rate the service first.'); return;
    }
    var selPercent = Number( eval( fldTipPercent.value));
    var billAmount = Number( eval( fldBillTotal.value));
    var tipAmount = ((selPercent / 100) * billAmount);
    var finalP = tipAmount + billAmount;
    fldTipRec.value = '$' + tipAmount.toFixed(2);
    fldFinalTotal.value = '$' + finalP.toFixed(2);
}
</script>
</head>
<body>
    <div class="toolbar">
        <h1 id="pageTitle">The Tipster</h1>
        <a id="backButton" class="button" href="#"></a>
    </div>
     <div id="main" title="Tipster" class="panel" selected="true">
        <h2 class="tip">Let the Tipster ease your pain and calculate the tip for
        you.</h2>
        <fieldset>
            <div class="row">
                <label>Bill amount:</label>
                <input type="text" id="fldBillTotal" value="20.00" tabindex="1"
                onfocus="clearTotal(this)" onchange="checkTotal(this)"/>
            </div>
            <div class="row">
                <label>Rating:</label>
                                <select id="fldTipPercent" onchange="getRec()"
                                        tabindex="2">
                            <option value="0">(Rate service)</option>
                            <option value="10">Very poor</option>
                                <option value="12.5">Poor</option>
```

```
                              <option value="15">Just as
                                      expected</option>
                              <option value="17.5">Above
                                      average</option>
                              <option value="20">Exceptional</option>
                              <option value="25">Wow!</option>
                          </select>
                </div>
            </fieldset>
            <fieldset>
                <div class="row">
                    <label>Tip: </label>
                    <input type="text" id="fldTipRec" value="0.00" readonly="true"
                    disabled="true"/>
                </div>
                <div class="row">
                    <label>Final total:</label>
                    <input type="text" id="fldFinalTotal" value="0.00" readonly="true"
                    disabled="true"/>
                </div>
            </fieldset>
        </div>
    </body>
</html>
```

The `fldBillTotal` input field captures the total before the tip. The `fldTipPercent` select list displays a set of ratings for the service, each corresponding to a percentage value (see Figure 14-6). These two factors are then calculated together to generate the output values in the `fldTipRec` and `fldFinalTotal` input fields.

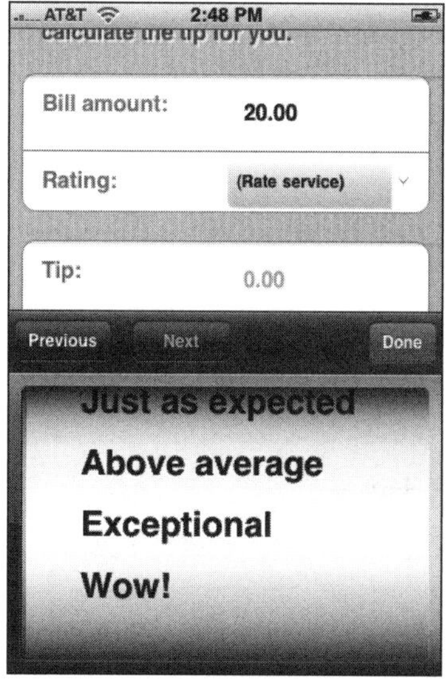

Figure 14-6: Scrolling through the select list

Next, I need to define the style rules for the mini app's UI. For core iPhone UI styling, I will use a subset of styles from a legacy version of iUI. I will add these inside a `style` element in the document head:

```css
<style type="text/css" media="screen">
body
{
    margin: 0;
    font-family: Helvetica;
    background: #FFFFFF;
    color: #000000;
    overflow-x: hidden;
    -webkit-user-select: none;
    -webkit-text-size-adjust: none;
}
body > .toolbar
{
    box-sizing: border-box;
    -moz-box-sizing: border-box;
    border-bottom: 1px solid #2d3642;
    border-top: 1px solid #6d84a2;
    padding: 10px;
    height: 45px;
    background: url(toolbar.png) #6d84a2 repeat-x;
}
.toolbar > h1
{
    position: absolute;
    overflow: hidden;
    left: 50%;
    margin: 1px 0 0 -75px;
    height: 45px;
    font-size: 20px;
    width: 150px;
    font-weight: bold;
    text-shadow: rgba(0, 0, 0, 0.4) 0px -1px 0;
    text-align: center;
    text-overflow: ellipsis;
    white-space: nowrap;
    color: #FFFFFF;
}
input
{
    box-sizing: border-box;
    width: 100%;
    margin: 8px 0 0 0;
    padding: 6px 6px 6px 44px;
    font-size: 16px;
    font-weight: normal;
}
body > .panel
{
    box-sizing: border-box;
```

```css
        padding: 10px;
        background: #c8c8c8 url(pinstripes.png);
    }
    .panel > fieldset
    {
        position: relative;
        margin: 0 0 20px 0;
        padding: 0;
        background: #FFFFFF;
        -webkit-border-radius: 10px;
        border: 1px solid #999999;
        text-align: right;
        font-size: 16px;
    }
    .row
    {
        position: relative;
        min-height: 42px;
        border-bottom: 1px solid #999999;
        -webkit-border-radius: 0;
        text-align: right;
    }
    fieldset > .row:last-child
    {
        border-bottom: none !important;
    }
    .row > input
    {
        box-sizing: border-box;
        margin: 0;
        border: none;
        padding: 12px 10px 0 110px;
        height: 42px;
        background: none;
    }
    .row > label
    {
        position: absolute;
        margin: 0 0 0 14px;
        line-height: 42px;
        font-weight: bold;
    }
    .panel > h2
    {
        margin: 0 0 8px 14px;
        font-size: inherit;
        font-weight: bold;
        color: #4d4d70;
        text-shadow: rgba(255, 255, 255, 0.75) 2px 2px 0;
    }
</style>
```

In addition to core iPhone UI styling, I need to define several app-specific style rules. `tip` classes are defined for the `h2`, `label`, `input`, and `select` elements. A separate `style` element is added to the document head to contain these styles:

```css
<style type="text/css" media="screen">
h2.tip
{
    margin-top: 10px;
    margin-bottom: 20px;
}
.row > label.tip
{
    position: absolute;
    margin: 0 0 0 14px;
    line-height: 42px;
    font-weight: bold;
    color: #7388a5;
}
.row > input.tip
{
    display: block;
    margin: 0;
    border: none;
    padding: 12px 10px 0 160px;
    text-align: left;
    font-weight: bold;
    text-decoration: inherit;
    height: 42px;
    color: inherit;
    box-sizing: border-box;
}
.row > select.tip
{
    display: inline;
    text-align: left;
    font-weight: bold;
    font-size: 12px;
    text-decoration: inherit;
    height: 36px;
    color: inherit;
    border: none;
    padding: 12px 0 0 10px;
    float: none;
    position: absolute;
    left: 150px;
    top: 3px;
    width: 140px;
}
</style>
```

Encoding Images

Although you now have all the styles and scripting code inside the HTML document, there is one last issue. Two of the styles reference external images for backgrounds. To use them, you need to encode these images first. The easiest way to do this is to use an online converter, such as the data: URI Image Encoder available at `www.scalora.org/projects/uriencoder`. This service performs a base64 encoding of a local file or a URL. You can then replace the image file reference with the attached encoded string:

```
body > .toolbar
{
    box-sizing: border-box;
    -moz-box-sizing: border-box;
    border-bottom: 1px solid #2d3642;
    border-top: 1px solid #6d84a2;
    padding: 10px;
    height: 45px;
    background: url(
"data:image/png;base64,iVBORw0KGgoAAAANSUhEUgAAAAEAAAArCAIAAAA2QHWOAAAAGXRFWHRTb2Z0
d2FyZQBBZG9iZSBJbWFnZVJlYWR5cc1lPAAAAE1JREFUCNddjDEOgEAQAgn//5qltYWFnb1GB4vdSy4WBAY
StKyb9+O0FJMYyjMyMWCC35lJM71r6vF1P07/1FSfPx6ZxNLcy1HtihzpA/RWcOj0zlDhAAAAAE1FTkSuQm
CC"
    ) #6d84a2 repeat-x;
}
body > .panel
{
    box-sizing: border-box;
    padding: 10px;
    background: #c8c8c8
url("data:image/png;base64,iVBORw0KGgoAAAANSUhEUgAAAAcAAAABCAIAAACdaSOZAAAAGXRFWHRT
b2Z0d2FyZQBBZG9iZSBJbWFnZVJlYWR5cc1lPAAAABdJREFUeNpiPHrmCgMC/GNjYwNSAAEGADdNA3dnzPl
QAAAAAE1FTkSuQmCC");
}
```

Now that all external resources are embedded, the application is fully standalone. However, you are not there yet. You now need to get it into a form that is accessible when the browser is offline.

Converting Your Application to a Data URL

You are now ready to convert your Web application into an encoded URL. Fortunately, several free tools can automate this process for you:

- *The data: URI Kitchen* (`software.hixie.ch/utilities/cgi/data/data`): This is probably the best-known encoder on the Web. It converts source code, a URL, or a local file to a data URL.

- *Url2iphone* (`www.somewhere.com/url2iphone.html`): This enables you to convert a URL into a bookmark. The most powerful aspect of this tool is that it looks for images, style sheets, and other files that are referenced and encodes these as well.

- *data: URI image encoder* (`www.scalora.org/projects/uriencoder`): This tool is great for encoding images into base64 format. You can specify a URL or upload a local file (see Figure 14-7).

- *Filemark Maker* (`http://mac.softpedia.com/get/Utilities/Filemark-Maker.shtml`): This is a free Mac-based utility that is oriented toward storing Word, Excel, and PDF documents as data URLs. However, it can also be used for HTML pages.

❑ *Encoding bookmarklet*: Developer David Lindquist developed a handy bookmarklet that grabs the current page's source, generates a data: URL, and loads the URL. You can then drag the generated URL onto your Bookmarks bar. Here's the JavaScript code:

```
javascript:x=new XMLHttpRequest();x.onreadystatechange=function(){
if(x.readyState==4)location='data:text/html;charset=utf-
8;base64,'+btoa(x.responseText)};x.open('GET',location);x.send('');
```

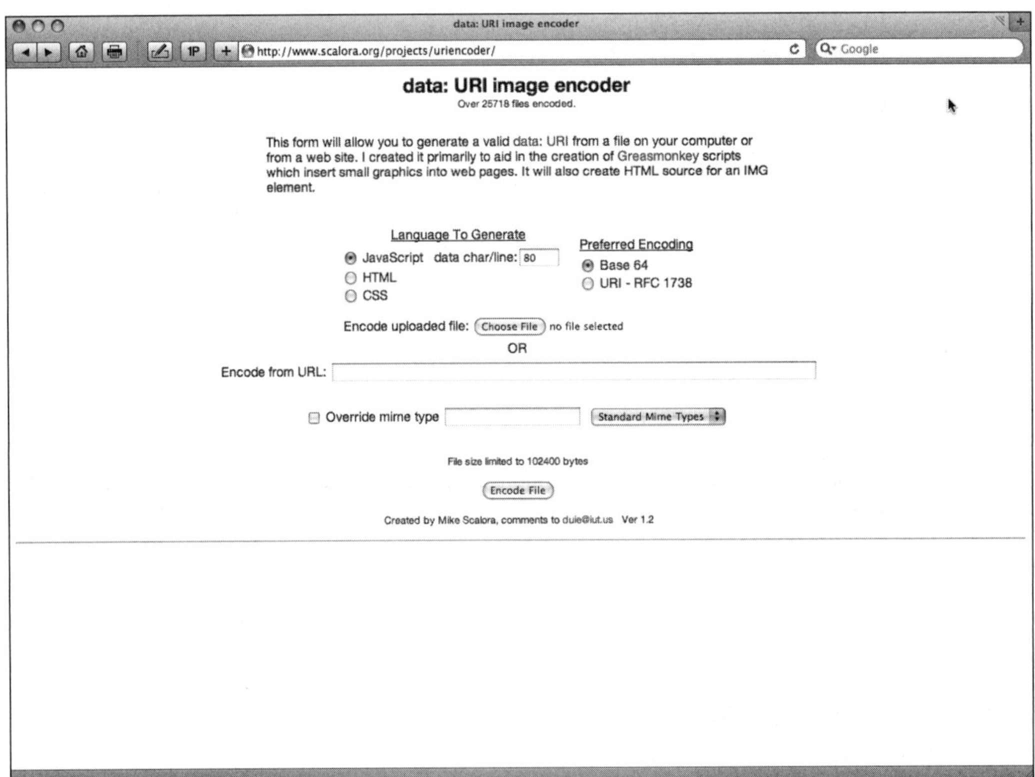

Figure 14-7: Encoding a Web application

❑ *Perl*: The following Perl syntax turns HTML into a data URL:

```
perl -0777 -e 'use MIME::Base64; $text = <>; $text = encode_base64($text);
$text =~ s/\s+//g; print "data:text/html;charset=utf-8;base64,$text\n";'
```

❑ *PHP*: In PHP, you can create a function to do the same thing:

```
<?php
function data_url($file)
{
  $contents = file_get_contents($file);
  $base64   = base64_encode($contents);
  return ('data:text/html;charset=utf-8;base64,' . $base64);
}
?>
```

Once you have used one of these tools to create a data: URL, make sure it is in the Address bar of Safari. Then drag the URL onto your Bookmarks bar. Synch up with your iPhone, and your application is now ready to run in offline mode. Figure 14-8 shows a fully functional Tipster.

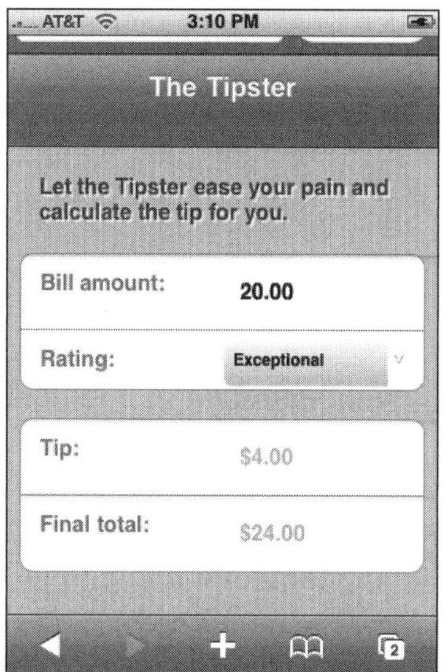

Figure 14-8: The Tipster application

Summary

In this chapter, you discovered how to work with bookmarklets and data URLs to create a special breed of offline iPhone web application. Bookmarklets are mini JavaScript scripts that can be stored as a book-marks inside of Safari on iPhone. A bookmarklet, which uses the `javascript:` protocol followed by script code, is typically used as a mini app that performs a very specific task or performs an action on the current Web page. The primary usefulness of bookmarklets to the iPhone Web application devel-oper is turning JavaScript into a sort of macro language to extend the functionality of Safari.

A data URL is a technique you can use for storing a Web page or complete application inside of a URL, which can then be saved as an iPhone bookmark inside Safari. The `data:` protocol allows you to encode an entire page's content inside a single URL. You can use data URLs to package Web applications and get around the requirement of a live 3G or Wi-Fi connection.

15

Debug and Deploy

Get in, get out. That's the attitude that most developers have in testing and debugging their applications. Few developers look forward to these tasks during the development cycle; they want to efficiently get into the code, figure out what's working and what's not, fix any problems, and then move on.

Given the heterogeneous nature of Web applications, debugging has always been challenging, particularly when trying to work with client-side JavaScript. To address this need, fairly sophisticated debugging tools have emerged over the past few years among the developer community, most notably Firebug and other add-ons to Firefox. However, the problem is that most of these testing tools that Web developers have come to rely on for desktop browsers are not ideal for testing iPhone Web apps.

Many iPhone Web app developers, unsure of where else to turn, are tempted to resort to `alert()` debugging — you know, adding `alert()` throughout the body of the script code to determine programmatic flow and variable values. However, not only is this type of debugging painful, it can throw off the timing of your script, making it difficult or impossible to simulate real-world results. Although the number of debugging and testing tools is indeed limited right now for Safari on iPhone, you still have options that either work directly inside Safari or emulate it on your desktop. You will probably want to incorporate aspects of both as part of your regular debugging and testing process.

Simulating the iPhone on Your Development Computer

Because you are coding your iPhone Web app on a desktop computer, you'll find it helpful to also be able to perform initial testing on the desktop. To do so, you'll want to create a test browser environment that is as close as possible to the iPhone. iPhone emulation enables you to more easily design and tweak the UI as well as test to see how your Web application or site responds when it identifies the browser as Safari on iPhone.

If you are running Mac OS X, you're in luck because you have three strong options. However, if you are developing on Windows, your options are more limited.

The SDK's iPhone Simulator

Your best option for simulating an iPhone on your desktop is to use the iPhone Simulator that is included with the Apple iPhone SDK. (The iPhone SDK is Mac OS X only.) Not only does this simulate the iPhone for testing native apps, but it has a built-in WebKit browser (see Figure 15-1) that simulates Safari on iPhone.

Figure 15-1: iPhone Simulator has a Safari emulator inside of it.

To use, simply click the Safari icon with your mouse. You can load your app from a bookmark or type a URL into the Address box (see Figure 15-2).

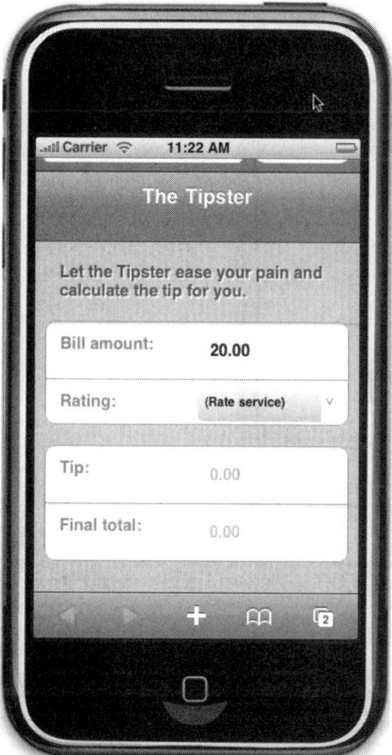

Figure 15-2: Loading an app inside the Simulator

The iPhone Simulator is actually quite powerful. Not only do the controls inside the Safari browser work as expected, but you can simulate rotation from the Rotate Left and Rotate Right items on the Hardware menu (see Figure 15-3). You can then test your app in horizontal mode (see Figure 15-4).

iPhoney

If you don't want to download the full Apple iPhone Developer SDK to use its iPhone Simulator, you can use iPhoney, a free open source tool created by Marketcircle (www.marketcircle.com/iphoney).

iPhoney (see Figure 15-5) can be a valuable tool to use when you are initially designing an iPhone Web app UI as well as when you are performing early testing. One of the handy features of iPhoney is that you can easily change orientations between portrait and landscape (see Figure 15-6). iPhoney also allows you to spoof with the iPhone user agent, hide the URL bar, and turn off Flash and other add-ins.

Figure 15-3: Simulating a rotation

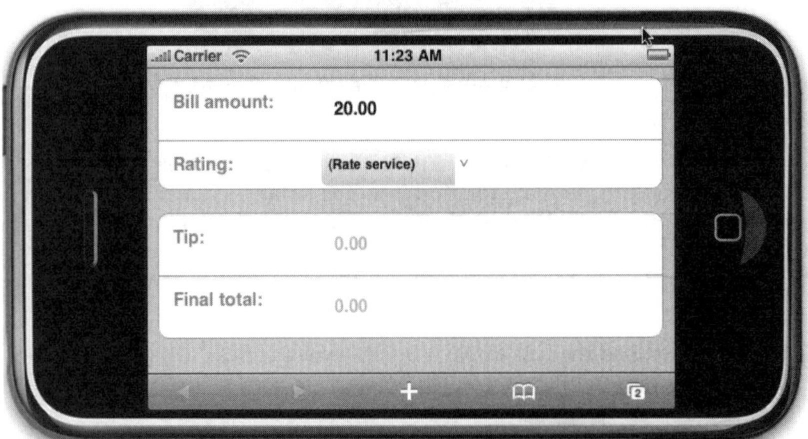

Figure 15-4: Testing the app in horizontal mode

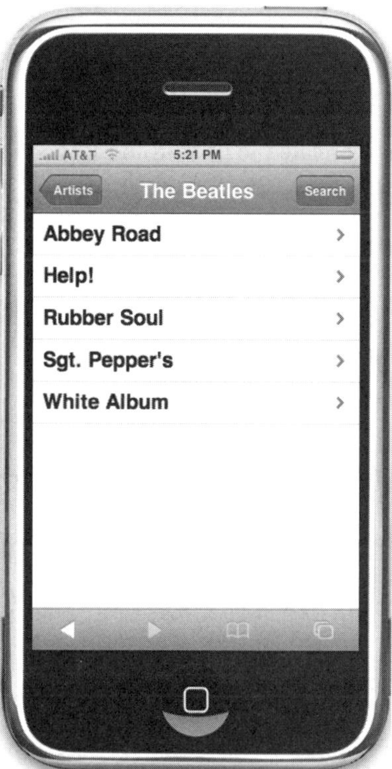

Figure 15-5: iPhoney simulates the iPhone on your Mac desktop.

Figure 15-6: Rotating iPhoney to landscape mode

Note that iPhoney requires Mac OS X.

Using Safari for Mac or Windows

Because Safari on iPhone is closely related to its Mac and Windows desktop counterparts, you can forgo these other custom tools and perform initial testing and debugging right on your desktop. However, before doing so, you will want to turn Safari into an iPhone simulator by performing two actions — changing the user agent string and resizing the browser window.

Changing Safari's User Agent String

Safari on Mac and Windows allows you to set the user agent provided by the browser through the User Agent list, which is accessible from the Develop menu.

Note: If you don't have the Develop menu available from the top menu bar, go to the Preferences dialog box and click the Advanced button. You'll see a check box there to enable the Develop menu (see Figure 15-7).

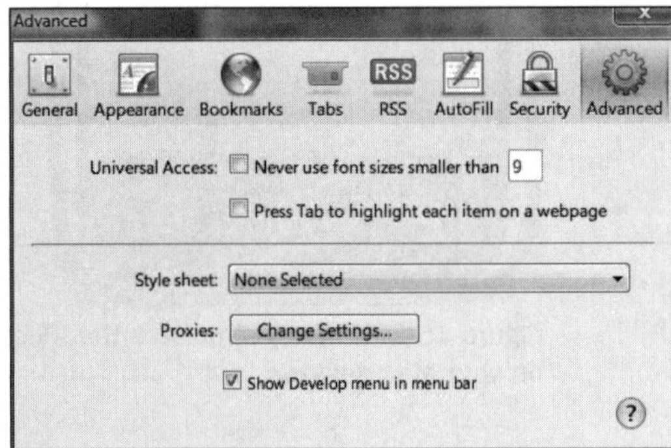

Figure 15-7: Enabling the Safari Develop menu

With the Develop menu enabled, choose the desired iPhone user agent string you want to emulate in from the User Agent menu, as shown in Figure 15-8.

After you have selected an iPhone user agent, the desktop version of Safari appears to your Web app as Safari on iPhone.

Changing the Window Size

To get the same viewport dimensions in Safari, you need to create a bookmarklet (see Chapter 14, "Packaging Apps as Bookmarks: Bookmarklets and Data URLs") and then add it to your Bookmarks bar. The code for the bookmarklet is as follows:

```
javascript:window.resizeTo(320,480)
```

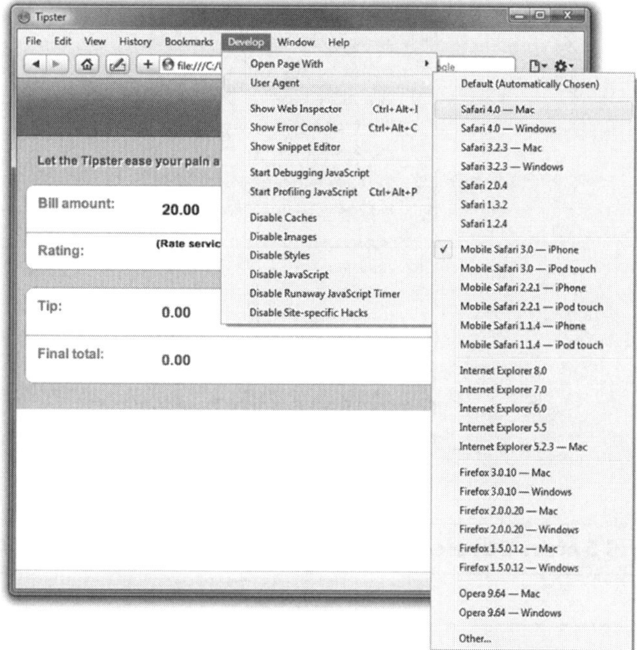

Figure 15-8: Choosing a Safari on iPhone user agent

Working with Desktop Safari Debugging Tools

Firefox has often been considered the browser of choice for Web application developers because of its support for third-party tools and add-ons, such as Firebug. However, when creating an application specifically for iPhone, it's usually best to work with Safari-specific tools. Fortunately, because Safari is so closely related to recent desktop versions of Safari, you can take advantage of the WebKit debugging tools that are provided with Safari for Windows and Mac. Because you are working with a close relative to Safari on iPhone, you still need to perform a second round of testing and debugging on an iPhone and iPod touch, but these tools can help you during initial Safari testing.

Working with the Develop Menu

The Safari debug tools are accessible through the Develop menu (see Figure 15-9), which is hidden by default when you install Safari. (As I mentioned in the previous section, you can enable it from the Advanced section of the Preferences dialog box.)

Many of the menu items in the Develop menu are not relevant to iPhone Web app development, but a few are worth mentioning (see Table 15-1).

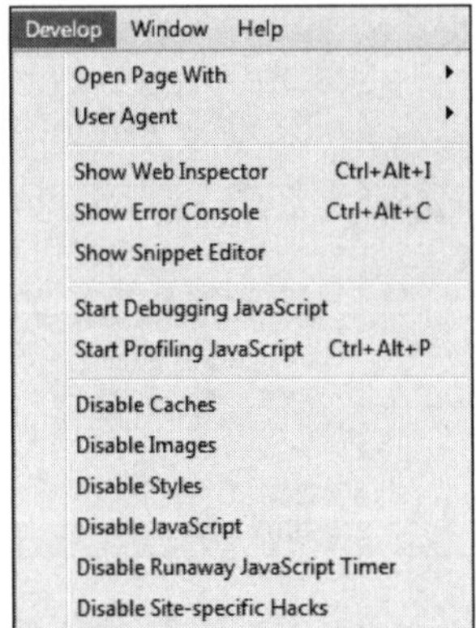

Figure 15-9: Safari's Develop menu

Table 15-1: Useful Safari Develop Commands for the iPhone Web App Developer

Name	Description
User Agent	Spoof another browser, particularly Safari on iPhone.
Show Web Inspector	View and search the DOM and styles.
Show Error Console	View error and status info in the Web Inspector.
Show Snippet Editor	Get instant rendering of an HTML snippet.
Log JavaScript Exceptions	Turn on to log exceptions.
Show JavaScript Console	View JavaScript errors occurring on a page.
Enable Runaway JavaScript Timer	Toggle the timer that halts long-running scripts.

The two Safari developer features worth special attention are the Web Inspector and JavaScript Console.

Working with the Safari Web Inspector

The best debugging feature available in Safari is certainly the Web Inspector. The Web Inspector, shown in Figure 15-10, enables you to browse and inspect the DOM of the current Web page. You can access this feature through the Develop menu. However, the handiest way to use it is to right-click an element in your document and choose the Inspect Element menu item. The Web Inspector is displayed, showing the element in the context that you selected in the browser window.

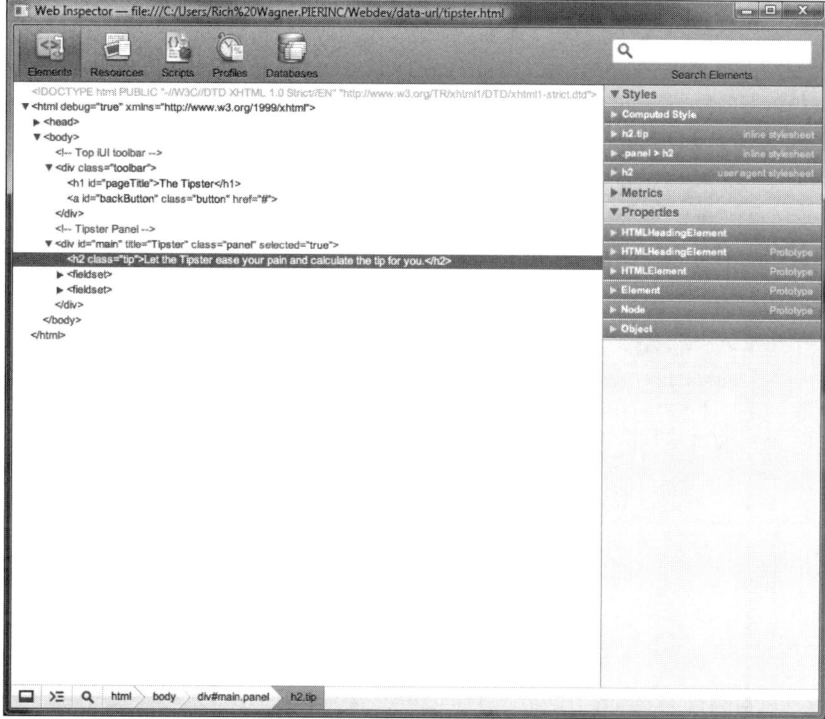

Figure 15-10: Web Inspector in Safari

Here are the basic functions of the Web Inspector:

- ❑ **Selecting a node to view:** When you click on a node in the Inspector pane, two things happen. First, the bottom pane displays node and attribute details, style hierarchy, style metrics, and property values. Second, if the selected node is a visible element in the browser window, the selected block is highlighted with a red border in Safari.

- ❑ **Changing the root:** To avoid messing with a massive nested DOM hierarchy, you can change the context of the Web Inspector. Double-clicking a node makes it the hierarchical "root" in the Inspector pane. Later, if you want to move back up the document hierarchy, use the up arrow or the drop-down combo box.

- ❑ **Searching the DOM:** You can use the Search box to look for any node of the DOM — element names, node attributes, even content. Results of the search are shown in the Inspector pane, displaying the line on which a match was found. If you want to get a better idea of the exact node you are working with, select it and then look for the red outlined box in the Safari window.

- ❑ **Viewing node details:** The Node pane provides basic node information, including type, name, namespace, and attribute details.

❑ **Viewing CSS properties:** The Style pane displays CSS rules that are applied to the selected node (see Figure 15-11). It demonstrates the computed style of the selected element by showing you all the declarations that are used in determining the final style of rendering. The rules are lists in cascade order. Any properties that have been overridden are displayed with strikethrough text.

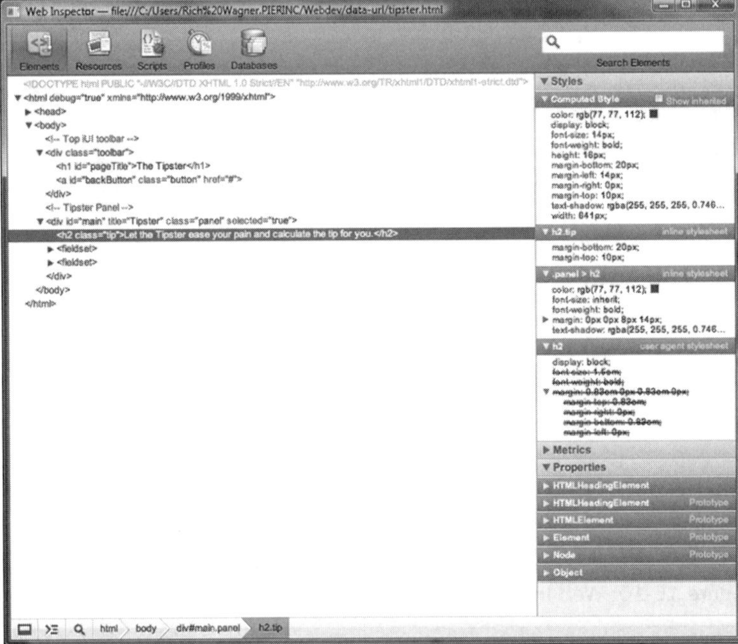

Figure 15-11: Style rules for the selected node

❑ **Viewing style metrics:** The Metrics pane presents the current element as a rectangular block displaying the width x height dimensions, as well as the padding and margin settings (see Figure 15-12).

❑ **Viewing all properties:** The Properties pane displays all the DOM properties (such as `id` and `innerHTML`) for the selected node. Because you cannot drill down on object types, this pane is less useful than the others.

Working with the Scripts Inspector

Safari's Web Inspector also sports a powerful Scripts Inspector, as shown in Figure 15-13. It's Safari's answer to `alert()` debugging: You can use it to inspect variables at point of execution, set breakpoints, step through your code, and view the call stack.

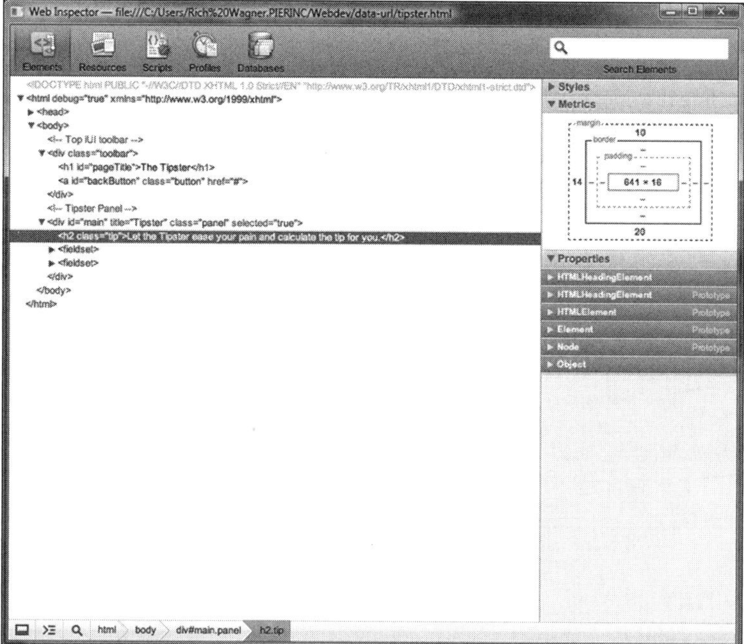

Figure 15-12: An element's metrics are easily seen in the Metrics pane.

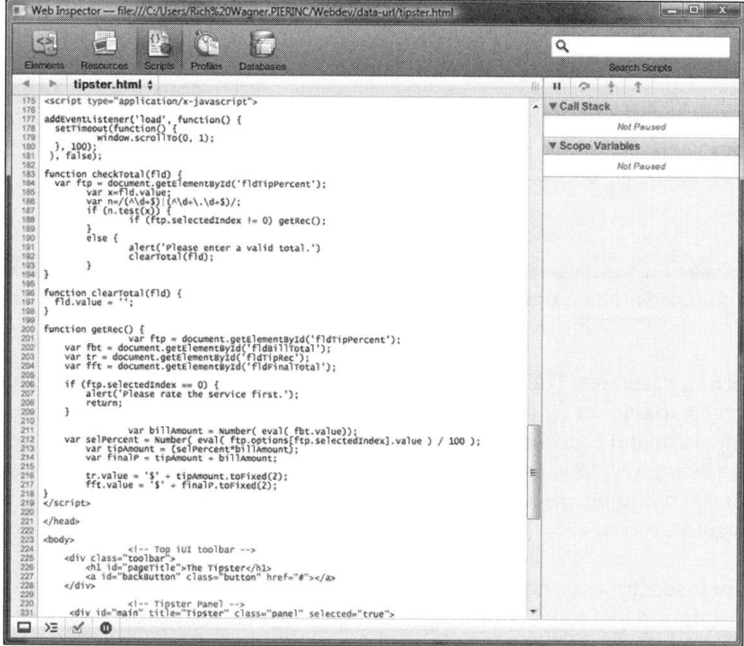

Figure 15-13: Safari's Script Inspector

You can perform several tasks in the Script Inspector:

❑ **Setting breakpoints and stepping through code:** You can set a breakpoint in your code by clicking the line number on the left margin of the code window. As Figure 15-14 shows, an arrow is displayed on the breakpoint line. When the line code is executed, the breakpoint is triggered. You can then step through the script as desired by clicking the Step Into, Step Out, and Step Over buttons. As you step through the code, Script Inspector updates its state for each line executed.

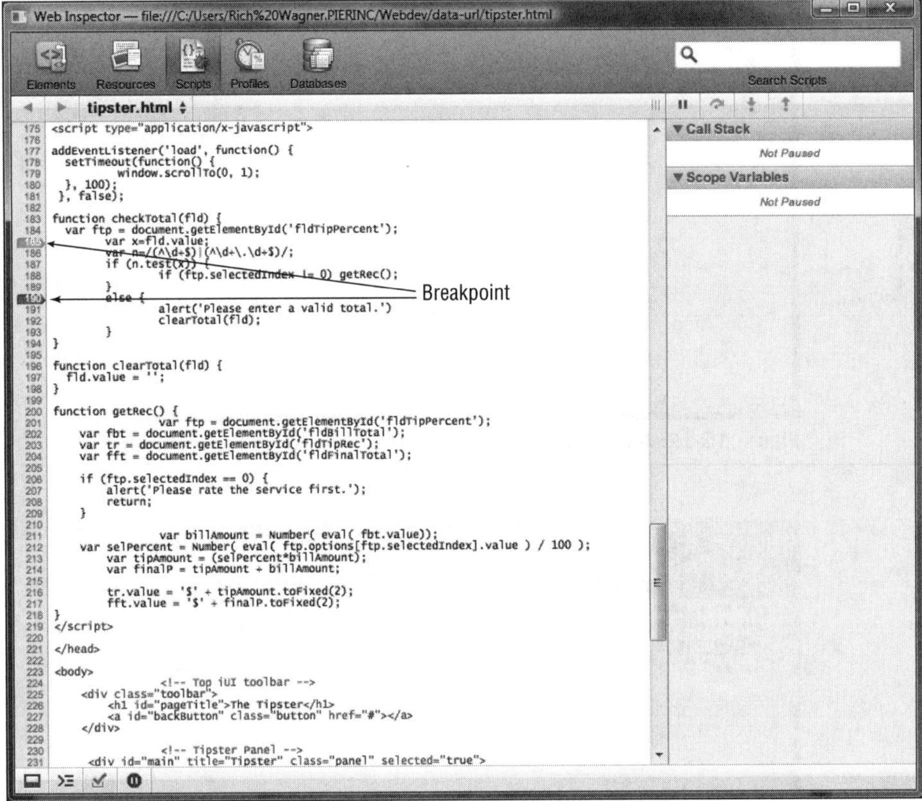

Figure 15-14: Setting a breakpoint

❑ **Inspecting variables:** The Scope Variables box at the top of the Script Inspector window displays the variables in scope (local, closure, global). You can inspect these variables by right-clicking them and choosing Inspect Element. A new Web Inspector window is displayed on top of the existing window, as shown in Figure 15-15, showing the node in its hierarchy along with style, metric, and property details. Close the Web Inspector window to return to the current debugging session.

Although Script Inspector does not work directly with Safari on iPhone, it does serve as the most powerful debugging option that the iPhone and iPod touch application developers have in their toolkit.

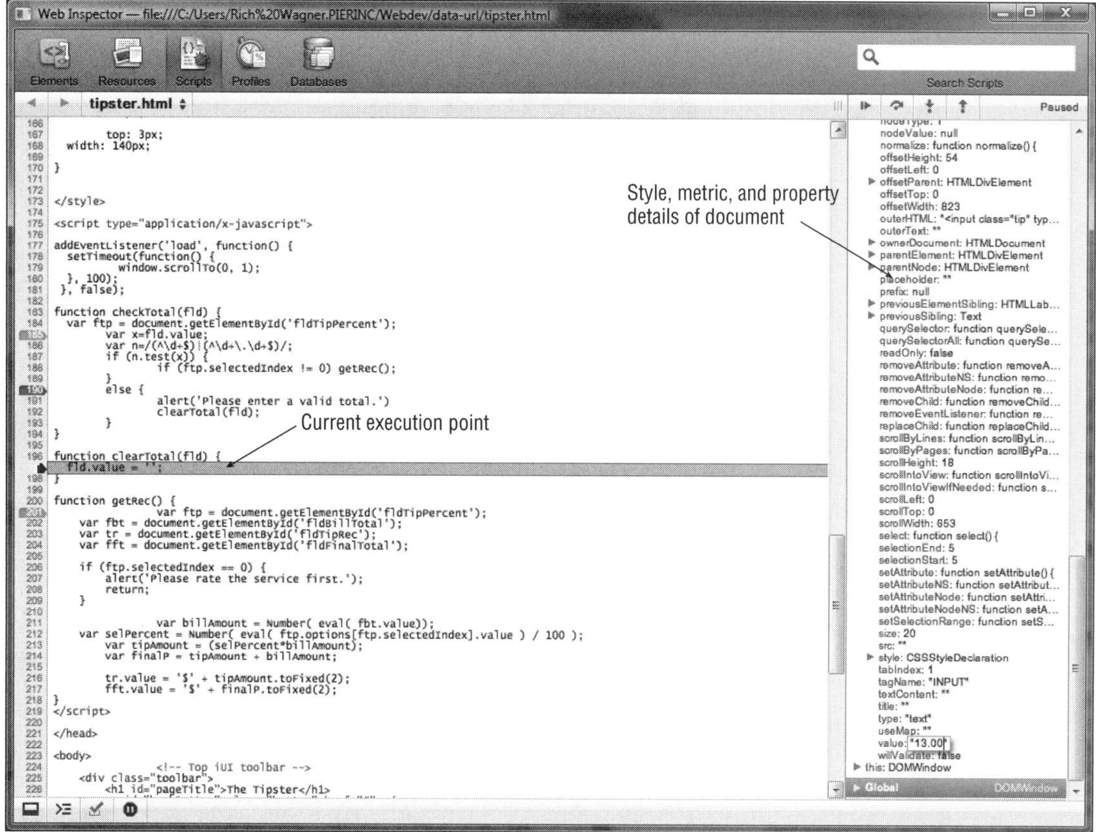

Figure 15-15: Inspecting the current state of an element in a debugging session

Debugging the iPhone

So far, you've seen how to test and debug your iPhone Web applications on your desktop using desktop-based solutions. Although those tools are good for general testing or specific problem solving, you may need to spend time debugging directly on the iPhone or iPod touch devices themselves. Unfortunately, no robust debugging tools such as Web Inspector are available, but some basic debugging tools are available that should be a standard part of your Mobile Safari development toolkit.

The Debug Console

Safari on iPhone includes an integrated Debug Console. If active, the Debug Console displays below the URL bar when an error is encountered or when it has a recommended tip (see Figure 15-16). You can click the right arrow to display a list of console messages (see Figure 15-17). The errors can be filtered by JavaScript, HTML, or CSS.

Figure 15-16: Debug bar displayed in Safari

Figure 15-17: Full Debug Console view

You can enable the Debug Console from the Settings app. Inside the Safari Settings, you can access a Developer. This panel enables you to turn on the Console with the slider control (see Figure 15-18).

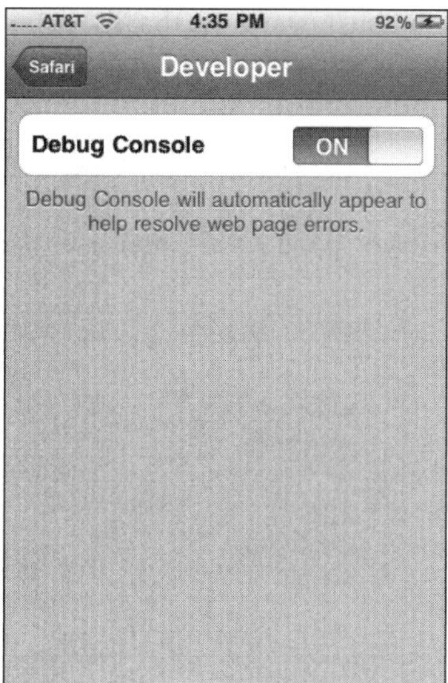

Figure 15-18: Enable the Debug Console

The DOM Viewer

The DOM Viewer, available from Brainjar.com, is a Web-based DOM viewer that you can work with directly inside of Safari on iPhone. The DOM Viewer provides an expandable tree structure that lists all the properties of a given node. When a property of a node is another node, you can view its properties by clicking its name. The tree expands to show these details. The DOM Viewer is housed in a separate HTML page that is launched in a separate window from the original page.

Although the DOM Viewer does not have the robust capabilities of the desktop Safari's Web Inspector, it does have the assurance that all the information you are looking at comes directly from iPhone's version of Safari, not its desktop cousins.

Starting the DOM Viewer

To use the DOM Viewer, follow these steps:

1. Download the source file at `brainjar.com/dhtml/domviewer.html`. Save the file in the same folder as your application.

2. Add a test link to your page to launch the viewer:

```
<a href="domviewer.html" target="_blank">View in DOM Viewer</a>
```

   Alternatively, you can add a script to the end of your HTML page that you want to inspect:

```
<script type="application/x-javascript">
window.open('domviewer.html');
</script>
```

The problem with this solution, however, is that iUI gets in the way of the default open action if you are using an iUI-based application.

3. Save the file.

4. Open the page inside Safari. If needed, click the View in DOM Viewer link.

The DOM Viewer is displayed in a new pane inside Safari (see Figure 15-19). Interact with it as desired.

Figure 15-19: DOM Viewer

Specifying a Root Node

One of the things you will immediately notice when working with the DOM Viewer inside the small iPhone viewport is the sheer amount of information you have to scroll through to find what you are looking for. To address this issue, the DOM Viewer allows you to specify a particular node (identified by id) as the document root. Here's the code to add, specifying the desired element as the getElementById() parameter:

```
<script type="application/x-javascript">
  var DOMViewerObj = document.getElementById("Jack_Armitage")
  var DOMViewerName = null;
</script>
```

Because it references the desired element directly by getElementById(), you can add this code in your HTML page *after* the element you want to examine in the body but not before it.

Go to brainjar.com/dhtml/domviewer for full details on the DOM Viewer.

Summary

Debugging is one of those necessary evils of any developer's experience, whether you are developing iPhone Web apps, native apps, or desktop widgets. In this chapter, I explored the variety of debugging tools and features available to iPhone Web app developers. I began by showing you how to simulate the iPhone on your desktop computer. I continued on the desktop by diving into Safari's developer tools. The chapter closed out by looking at the debug features that you can utilize on the iPhone itself.

16

The iPhone SDK: From Web App to Native App

As I discussed in Chapter 1, "Introducing Safari/WebKit Development for iPhone 3.0," the world of iPhone applications has two breeds: native apps that are installed and run on the device, and Web apps that run inside of Safari on iPhone. The focus of this book has been on creating Web-based applications. But suppose you develop a Web app and find that, as time goes on, needs change and you must consider a native solution instead. Or suppose your user base is clamoring for a native version of your app, but you have no clue how to go about developing one.

If your experience has been primarily with Web technologies, don't simply dismiss the idea of porting to a native solution because *you do Web*. You may find that moving your app to the native iPhone platform is not only the best move for your application and user base, but the best move for you professionally — learning a new skill set while you port.

In this final chapter, I'll introduce you to native iPhone application development. As part of that, you'll explore the advantages and disadvantages of both breeds as well as the iPhone 3.0 SDK. I'll also talk about PhoneGap, which is a unique open source solution for wrapping your Web-based app inside a native shell.

Comparing the Strengths of Web Apps and Native Apps

Let's face it: The iPhone is not a level playing field for native and Web apps. Unfortunately for Web developers, it is biased toward native solutions. The iPhone App Store only carries native apps, not Web-based ones. Apple does not have a way to allow users to "install" a Web app on their home screen automatically; instead, it requires the user to go through the manual step of adding just as you would a browser bookmark. What's more, until recently, Web apps could not fully customize the Safari chrome (or UI shell). Therefore, you might conclude that the native route is always the way to go. Not so. As you've seen throughout this book, in spite of the aforementioned handicaps, you can create compelling applications inside of Safari on iPhone in which you can provide a user experience every bit as solid as a native app.

Web App Advantages

Web apps have several distinct advantages over native apps:

- **Leverage existing know-how and expertise:** Perhaps the most personal and compelling advantage for Web developers is the fact that Web apps allow you to create iPhone solutions using your existing skill set and expertise. Instead of dealing with a learning curve, all you need to do is augment your existing knowledge base to deal with unique aspects of iPhone as a client.

- **Leverage existing technology:** A Web solution enables you to directly leverage existing Web-based solutions and simply develop a presentation layer that works well and intuitively as an iPhone app.

- **Lower learning curve:** Native applications are written in Objective-C, which is an object-oriented version of C. It is the ubiquitous programming language of Mac OS X development and is naturally extended to the iPhone. If you have never worked with a lower-level language like C or C++ before, it can be quite a jolt to the system. You can feel like you've stepped backward and have to deal with all sorts of issues that you simply don't think about when using a language such as JavaScript. There is no garbage collection, so you have to manage memory within your application; everything you create, you have to dispose of. You also have to concern yourself with challenging programming concepts like pointers and threads. What's more, although the iPhone SDK tools may look visual and advanced, the Objective-C code can look quite arcane. The code is nonlinear and is based on delegates, which are called only when the OS is ready. As a result, it's difficult at first to figure out the logical order and progression of an application.

  As a result, Web application development has a distinct advantage over native application development for existing web developers or for those with no formal programming background at all. Web technologies, such as JavaScript, Ajax, CSS, and HTML, have a much lower learning curve than Objective C does. You don't need to worry about such low-level issues as memory management, pointers, and threads. Instead, you can let Safari deal with those issues while you focus on your solution.

- **Instant distribution without bottlenecks:** If you keep up with tech news, you probably know already that one of the major handicaps that native apps have is the App Store bottleneck. It can take several weeks for Apple to approve a native iPhone app. What's more, this approval process is not just a one-time thing. Even if you are releasing a minor bug fix version, you can often experience the same processing and approval delays. In sharp contrast, because an iPhone Web app is a server-based solution, you can deploy an update and redistribute your app at a moment's notice. There are no barriers or bottlenecks to doing so.

- **Easy to maintain and update:** Not only do you not have distribution bottlenecks with a Web app, but a Web app is much easier to maintain and update because you control the program on the backend.

- **Near-native capabilities:** I argued before that the iPhone seems to have a bias toward native solutions over Web-based ones. At the same time, although the technology gap between the two was large during the initial release of the iPhone, this gap has largely disappeared. iPhone OS 3.0 gives Web developers access to the key touch events, the ability to fully customize the Safari chrome, and a way to get onto the home screen (even if the way in which it gets there is not optimal).

What's more, with HTML 5's local storage capabilities, Web apps no longer depend on a live connection with the Internet. Therefore, although Web apps may still not be fully on the same playing field as native apps, the gap continues to shrink.

Google Latitude, shown in Figure 16-1, is one of the best examples of a popular iPhone app that is Web based rather than native.

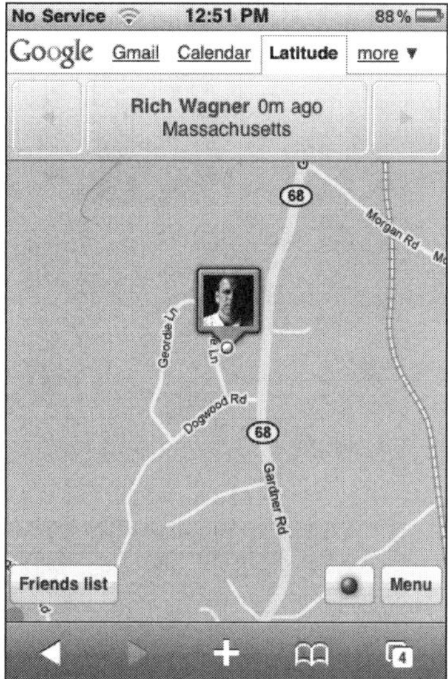

Figure 16-1: Google Latitude deployed as a Web app

Native App Advantages

Given the incredible popularity of the App Store, it is quite clear that a native app architecture has some advantages. Here are several of them:

❑ **Compelling user experience:** When you interact with a well-made native iPhone app, there is an undeniable "look and feel" factor. It has a snappiness and robustness that make it a joy to use (see Figure 16-2). You could sum it up with the following statement: *Web seeks to emulate, but native allows you to innovate.*

❑ **Programming power:** Although the innovations in Ajax, CSS animations, and HTML 5 give Web developers more and more power to create powerful applications, Web technologies will never provide the raw programming power needed for games and other processor-intensive programs (see Figure 16-3).

❑ **Speed:** Safari interprets the client-side source code base of Web apps at run time before their instructions are processed. In contrast, native iPhone apps are compiled. As a result, native apps always execute faster than Web apps. What's more, unless the native app accesses Web services, some or all of its data and resources are stored on the iPhone. In contrast, Web apps always have to deal with bandwidth and latency issues.

Figure 16-2: Newsstand RSS reader

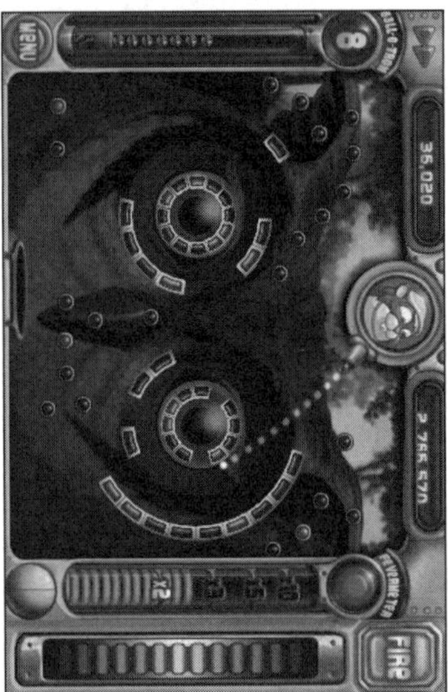

Figure 16-3: Paggle, an app that takes
advantage of the power of the iPhone

❑ **Access to iPhone hardware and software services:** Although Apple has made many of the services available to Web app developers, native apps get even more, such as the camera, compass, and photo gallery.

❑ **No Internet dependency:** Although Web developers can take advantage of HTML 5's offline cache to operate when the Internet is not available, Web apps tend to have an underlying dependency on the Internet. In contrast, because native apps are installed on the device, there is no inherent dependency on the Internet.

❑ **Income generation:** Apart from intra-app advertisements, there are no successful models for making money from iPhone Web apps. On the other hand, the App Store (see Figure 16-4) has emerged as the poster child of success stories for application developers. Therefore, if you are planning on trying to earn income from your work, a native app may be your only viable option.

Figure 16-4: The App Store has provided many "Rags to Riches" stories for application developers.

Knowing When It's Time to Port Your App

One of the great advantages of iPhone Web apps is the ability to get something up and running. And, then, once you have an initial version deployed, you are free to decide where to invest your time and energies for future releases. Do you continue to develop a Web-based solution? Or do you decide to port your Web app to the native platform?

Here are some issues to keep in mind during your Web vs. native decision-making process:

❑ **Nature of app:** Does the nature of the app lend itself to a native or Web model? Often, the key factor is whether the problem domain the app is in more or less makes the decision for you. If you are trying to create a game, the answer is simple: native gives you the graphics processing power you need. If you are creating a camera add-on, you'd go native to get the integration capabilities. On the other hand, if your app depends on Web services, a Web solution may be appropriate.

The choice is not always obvious. Google Latitude (see Figure 16-1) and the Amazon app (see Figure 16-5) are two apps that are completely reliant on Web services, yet choose opposite paths.

❑ **Existing technology:** Do you have existing Web-based technology that you want to continue to utilize? If so, you could access it in an API fashion using a native app (such as the Amazon app). Or you may decide that a Web app solution gives a sound user experience all the while making the most of your existing investment (see Figure 16-6).

❑ **Timing:** How important is time to market? A native app has a considerable learning curve for new developers, has a longer development cycle to develop comparable functionality, and is subject to the often-lengthy App Store review process.

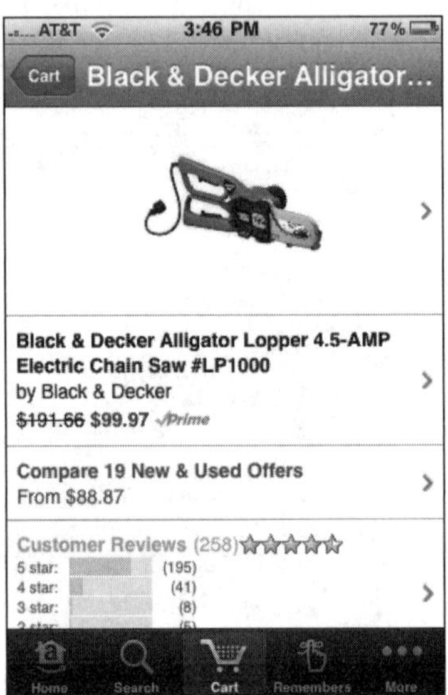

Figure 16-5: Amazon's native app relies entirely on Web services.

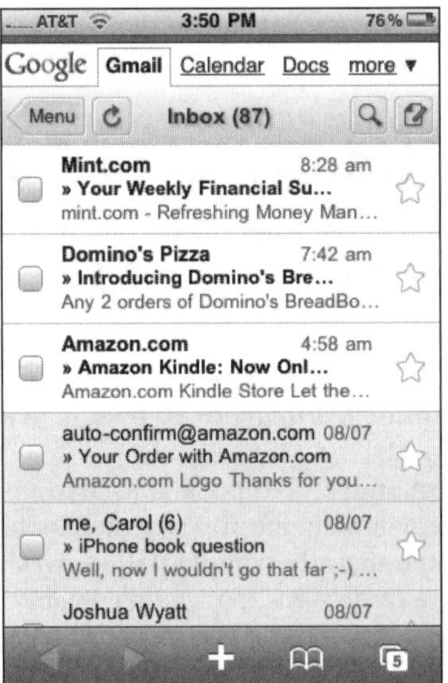

Figure 16-6: Google chooses the Web route for Gmail access on iPhone.

❑ **User feedback:** What are your users saying about their experience with your app? Do you get complaints about speed? Do you have iPod touch users who would like to use your app all the time, not just when they have Wifi access? Or do you have users frustrated that updates are too few and far between?

Peeking into the iPhone SDK

Native iPhone app development is centered on the iPhone SDK. The SDK contains all the tools and code you need to create native iPhone apps. It's available for Mac OS X only, so if you are a Windows developer, you're out of luck.

The SDK Process

The basic process for developing and deploying an iPhone app is as follows:

1. Create a developer account at `http://developer.apple.com/iphone`.

2. Download the free iPhone SDK (Mac only).

3. Install the iPhone SDK, which includes the Xcode development tool.

4. Develop your app using Xcode and its suite of SDK tools.

5. Preview and debug using the Mac-based iPhone Simulator.

6. If you want to actually install your app on an iPhone device, you need to make a one-time purchase of an iPhone Developer Program certificate ($99).

7. Once approved, you can submit an application to the App Store (or more specifically the online Developer Program Portal).

iPhone SDK Tools

The iPhone SDK consists of a suite of tools and other resources called Xcode that you use to create a native Objective C app. Here is an overview of the pieces of the toolkit that you'll work with first:

❑ **Xcode IDE** is your primary tool for writing and editing source code and managing your project (see Figure 16-7).

❑ **Interface Builder** is a visual tool used to design your application's UI. The UI you create is then loaded by the app as resources (see Figure 16-8). You can create your own design from scratch or use UIKit templates to give you a head start.

❑ **iPhone Simulator** (which I discuss in Chapter 15, "Debug and Deploy") is used to preview and debug your apps on your desktop machine. It looks like the iPhone, except on your desktop, that is (see Figure 16-9).

❑ **SQLite** comes as part of the SDK. If you need a database for your app, you can utilize the SQLite database management library.

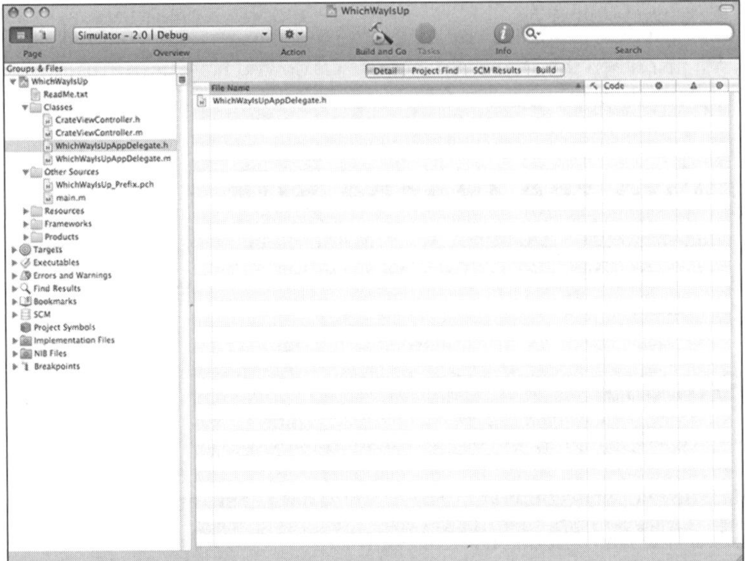

Figure 16-7: Xcode IDE

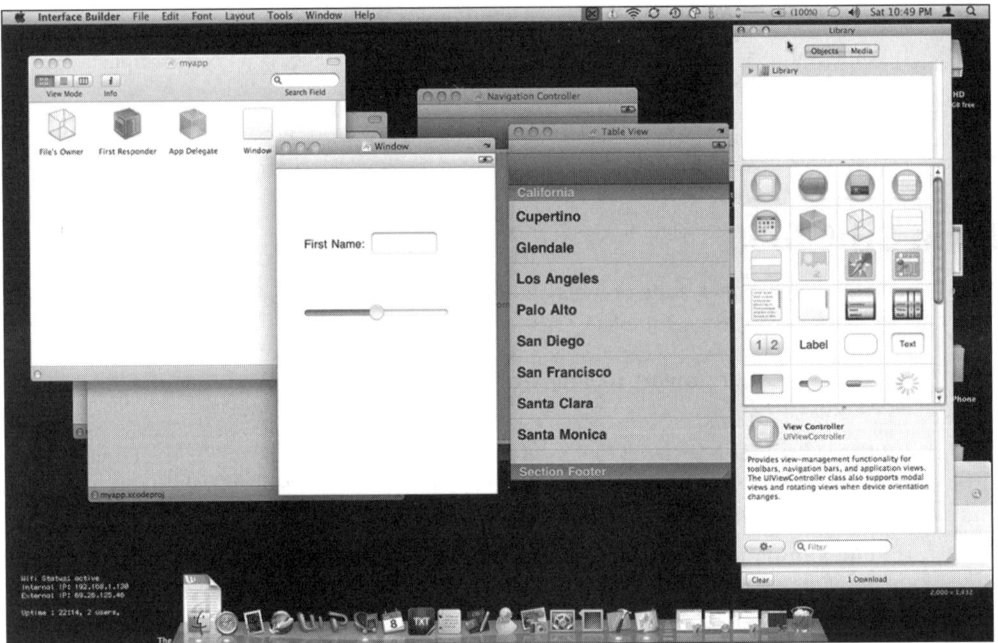

Figure 16-8: Design your UI in Interface Builder

Figure 16-9: iPhone Simulator allows you to test your app on your desktop.

PhoneGap: A Hybrid Solution

PhoneGap (`http://phonegap.com`) is an open source development tool that offers an interesting hybrid model to consider. You can build your application using familiar Web technologies but wrap it in a native iPhone shell. In other words, your Web app is packaged with a customized native WebKit browser that can be added to the App Store, alongside other native apps. Its developers express that the purpose of PhoneGap is "to solve device integration by Web-enabling devices' native functionality with open standards." What's more, PhoneGap enables you to do the same things for other mobile platforms, including Android and Blackberry.

PhoneGap enables you to create a native shell for your Web app using PhoneGap along with the iPhone SDK. After downloading the latest source from the PhoneGap Web site, you can start off by opening the `PhoneGap.xcodeproj` file, inside the `iphone` subdirectory (see Figure 16-10). The project is displayed in the Xcode window (see Figure 16-11).

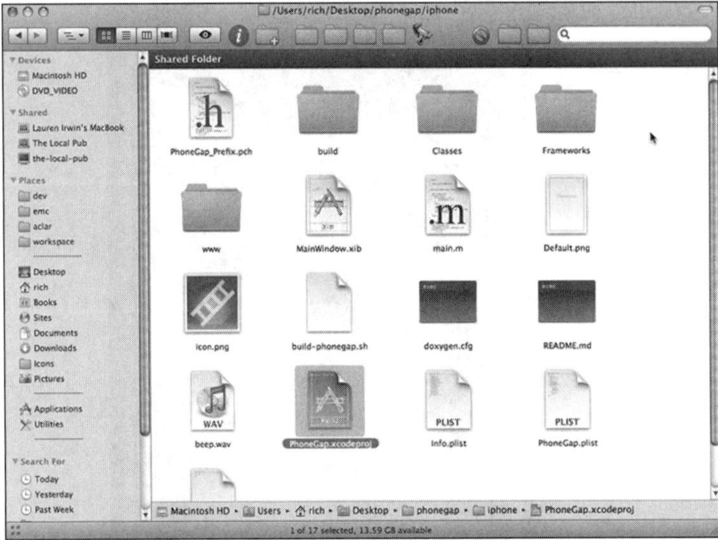

Figure 16-10: Opening the Xcode project file

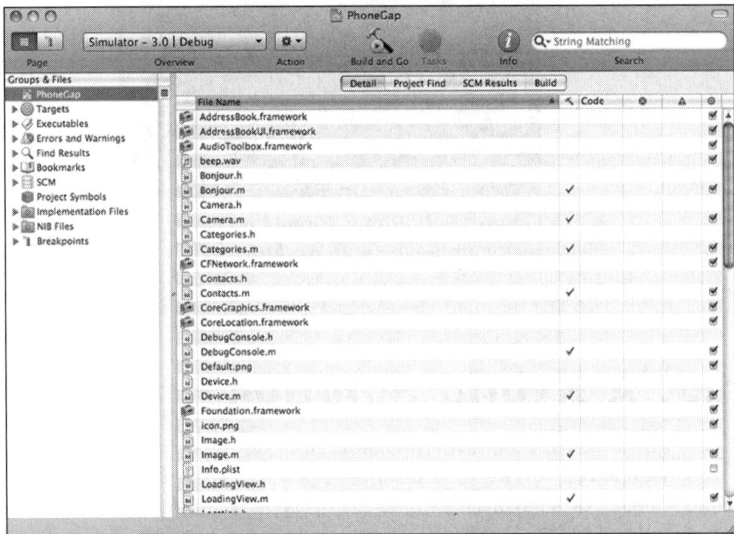

Figure 16-11: Xcode window

In the `iphone/www` subdirectory, you'll find an `index.html` file (see Figure 16-12). The app displays this page when it is run. You can replace the existing file with your own Web app starting page, including necessary resources and other HTML pages in the www folder.

When you run the app in the simulator, the PhoneGap project provides a shell to the Web page you provided (see Figure 16-13).

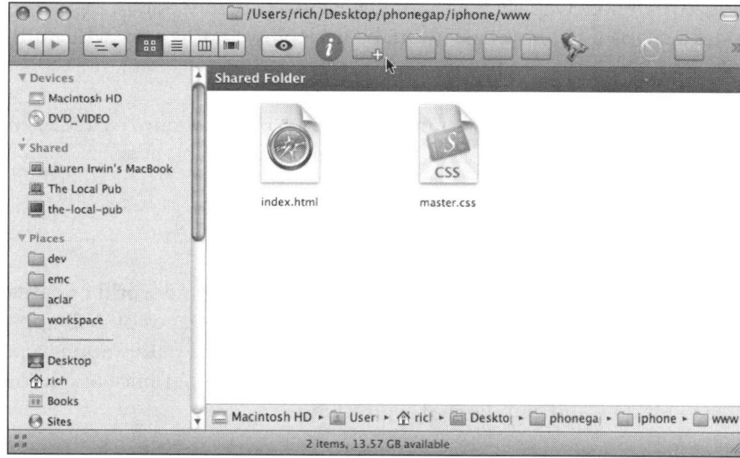

Figure 16-12: Locating the `index.html` page

Figure 16-13: Web page used as the main
UI of the PhoneGap app

You can customize the PhoneGap shell and undertake tasks before deploying your app, but this simple example provides a snapshot of how easy it can be to deploy a native app while utilizing your existing Web technology.

You can download the latest PhoneGap source at `http://github.com/sintaxi/phonegap/`.

Summary

In this final chapter, you were introduced to the iPhone SDK and the world of native application development. I compared and contrasted the advantages and disadvantages of Web apps and native apps. In the end, whether you should consider migrating a Web app to a native app is a personal or team decision, based on a variety of factors including programming experience and know-how, desired user experience, time to market, and commercial aspirations.

Then, after a survey of the key tools of the iPhone SDK, the chapter finished with a look at how you could distribute your Web app in a native shell using PhoneGap.

Index